Practical adjudication
for construction professionals

Leslie J. Edwards BSc FICE FCIArb FIRM
and
Richard N. M. Anderson Barrister

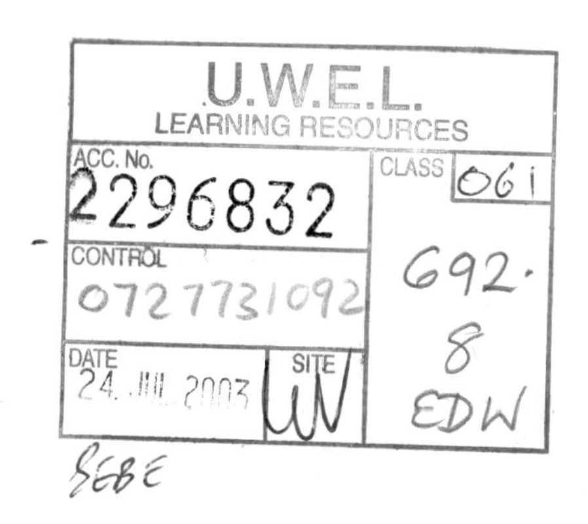

Thomas Telford

Published by Thomas Telford Publishing, Thomas Telford Ltd,
1 Heron Quay, London E14 4JD. URL: http://www.thomastelford.com

Distributors for Thomas Telford books are
USA: ASCE Press, 1801 Alexander Bell Drive, Reston, VA 20191-4400, USA
Japan: Maruzen Co. Ltd, Book Department, 3–10 Nihonbashi 2-chome, Chuo-ku, Tokyo 103
Australia: DA Books and Journals, 648 Whitehorse Road, Mitcham 3132, Victoria

First published 2002

Also available from Thomas Telford Books

The New Engineering Contract: a legal commentary. Arthur T. McInnis. ISBN 0 7277 2961 6
NEC and partnering. John Bennett and Andrew Baird. ISBN 0 7277 2955 1
Adjudication under the NEC. Richard N. M. Anderson. ISBN 0 7277 2997 7
Practical risk management in the construction industry. Leslie Edwards. ISBN 0 7277 2064 3
Civil engineering insurance and bonding. 2nd ed. Leslie Edwards, Geoffrey Lord and Peter Madge. ISBN 0 7277 2098 8
Professional services agreements. Leslie Edwards and Rachel Barnes. ISBN 0 7277 2884 9

A catalogue record for this book is available from the British Library

ISBN: 0 7277 3109 2

Designed and typeset by Bookcraft Ltd, Stroud, Gloucestershire.
Printed and bound in Great Britain by MPG Books Ltd, Bodmin, Cornwall.

Disclaimers

Generally

The publishers and authors expressly disclaim all and any liability and responsibility to any person, whether a purchaser of this work or otherwise, in respect of anything done or omitted to be done by any such person in reliance, in whole or in part, upon the contents of this work.

Worldwide

The above general disclaimer applies but, in addition, the attention of users worldwide is specifically drawn to the fact that either the law of the contract and/or the law of the country may have an impact upon the process of adjudication being used there and readers adjudicating there should accordingly seek the advice of a competent professional adviser.

United Kingdom

The above general disclaimer applies but, in addition, the attention of users in the United Kingdom is specifically drawn to the fact that the legislation applying to adjudication is relatively new. It has yet to be fully tested, drawn, and developed. Accordingly, although references to court decisions of first instance are made in this book, some of those decisions might later be changed on appeal or otherwise. In relation to any specific transactions readers should seek the advice of a competent professional.

Copyright acknowledgements

We reproduce with kind permission of the ICE and JCT extracts from their Adjudication Procedures.

We reproduce with kind permission of the Construction Umbrella Bodies the most recent edition (May 2002) of their Draft Guidance put out for consultation.

We reproduce under licence from HMSO certain selected parts of the GC Adjudication Procedures.

We reproduce with kind permission of the Construction Industry Council extracts from their Model Adjudication Procedures, copies of which can be obtained from the Construction Industry Council, 26 Store Street, London WC1E 7BT.

Dedication

This book is dedicated to Sir Michael Latham, the father of the rebirth of adjudication.

Contents

Part II The conduct of an adjudication

Acknowledgements

The assistance on early drafts of this book by Paul Carter of Kennedys, solicitors, on legal matters and Kole Ojo of Willis Corroon Construction Risks Limited, on insurance matters, is gratefully acknowledged.

Thanks are also due to Mouchel Consulting Limited who provided time for the research and preparation of this book, as well as Jane Wright, Sue Barlow and Linda Cook who undertook much of the typing and retyping.

Finally the authors would like to thank the staff of Thomas Telford, and Jeremy Brinton in particular, for their support in a period of constantly changing case law.

About the authors

Leslie Edwards is a Senior Consultant with Mouchel Consulting Limited, West Byfleet.

After varied UK and overseas consultancy experience, Leslie has spent most of the last fifteen years providing expert civil engineering technical, contractual and risk advice on a wide variety of construction issues for private, insurance market, solicitor and contractor clients. The scope has included contractual, insurance, technical, professional negligence, workmanship and personal injury issues. He is a trained Arbitrator, Mediator and Conciliator as well as being a practising Adjudicator.

He is the author for Thomas Telford of *Practical risk management in the construction industry* and the joint author of *Civil engineering insurance and bonding* and of *Professional services agreements.*

Richard N. M. Anderson is a specialist in construction law and a trained and accredited Mediator, Adjudicator and Arbitrator. He has in the past been seconded to a large international construction company and is now in practice as a Barrister. His interest in adjudication in particular has developed from over ten years' experience of sitting as chairman of a panel for one of the Government Adjudication Schemes. Richard has advised, lectured and written widely on construction and adjudication issues and is an occasional case editor in the *Construction Law Journal.*

He is also the author of Sweet and Maxwell's *A Practical Guide to Adjudication in Construction Matters, Adjudication under the NEC* for Thomas Telford and the *Construction Adjudication Casebook* for Butterworths.

Richard N. M. Anderson
Barrister
Arbitration Chambers
22 Willes Road
London
NW1 8NY
DX 46454 Kentish Town
Tel 020 7267 2137
Fax 020 7482 1018
e-mail: andersons@coacc.fsnet.co.uk

Preface

Practical Adjudication has been written by a practising Adjudicator and a Barrister to provide an explanation of, and detailed practical guidance on, the process of construction-related adjudication to those that may become involved in it, whether as Adjudicators, participating Parties, advisors or insurers.

The adjudication requirements of Part II of the Housing Grants, Construction and Regeneration Act 1996 [the 'Construction Act'], the procedures specified in The Scheme for Construction Contracts (England and Wales) [the 'Scheme'], the ICE, JCT and GC/Works adjudication procedures (which are three of the current main contractual standard adjudication procedures) and the CIC stand-alone adjudication procedures are each described and, where considered appropriate, critically reviewed.

Incorporating recent decided legal cases, the book identifies differences between statutory Scheme adjudications and contractual adjudications, as well as the increasing divergence in some areas between the decisions of the English and Scottish courts.

It is understood that the Government may be making legislative changes to the Scheme and possibly the Construction Act. It is not known when this might happen or the details of the changes. However, for interim purposes, included in the text are details of the draft proposed Scheme for Construction Contracts (England) (Amendment)Regulations 2001 and the Construction Umbrella Bodies Adjudication Task Group's 2 May 2002 draft document *Adjudication under the Scheme – Guidance to Adjudicators*, which is based on what is understood to be current thinking on the foregoing draft Scheme amendments.

The law is stated as at 12 April 2002.

Chapter 1

Introduction

A large body of law has developed over the years relating to construction contracts. When the Parties to a contract cannot agree on a difference that has arisen between them arising out of that contract, the difference can be resolved by arbitration or litigation. The difference will commonly relate to the performance of the contract (i.e. there has been a possible breach of contract by one of the Parties) or to a disagreement over what one Party considers is additional payment or time due as a consequence of an error, action or inaction by the other Party.

Most conditions of contract will incorporate clauses that permit, if not oblige, the Parties to resolve their differences by arbitration, which is a quasi-judicial procedure now governed by Acts of Parliament and precedent that results in a binding decision that the courts would be reluctant to overturn. Perhaps less frequently, a construction dispute will still be referred directly to litigation, usually the Technology and Construction Court.

Often conditions of contract will also incorporate one or more alternative dispute resolution (ADR) procedures such as conciliation, mediation, expert determination, Engineer's decisions and adjudication, which must be tried, or at least are available to be chosen, before the Parties go to arbitration or litigation. Individual ADR procedures may be defined differently in different conditions of contract and can have their own rules and procedures of varying complexity and detail. A common feature is that they involve a third, normally neutral, party to assist achieve settlement. They do not preclude the Parties from continuing to negotiate directly between themselves. Another feature of ADR procedures is that generally any settlement reached is not necessarily final and binding unless the Parties agree to make it so. This compulsion can be incorporated within the original conditions of contract, or the Parties can later agree between themselves that this will be the case.

It could be said that there are four main types of conditions of contract used in the UK construction industry. The first are national standard industry conditions produced by one, or more usually a combination of industry bodies, and used without significant contractual amendment. The second are the industry standard conditions, albeit used with amendments to either more accurately reflect actual project requirements and/or to allocate risk differently between the Parties. The third conditions are 'ad hoc' internal conditions produced by, and standard to, individual organisations and companies such as employers, contractors or professionals. The fourth are also ad hoc conditions but entirely one-off, resulting from attempting to provide a contractual basis for a project not naturally covered by the standard conditions and with the risks allocated in a way that the principal Parties are prepared to accept.

Most construction contracts will involve as Parties either an Employer and a main Contractor, main Contractor and sub-Contractor (manufacturer, supplier, professional, other Contractor), Employer and professional, or professional and sub-Consultant. These will normally use one of the four previously described types of conditions of contract.

The result is that different combinations of dispute resolution procedures can be found within different conditions of contract. For example, it is possible in one case to find a progression through Engineer's decision, conciliation, adjudication and arbitration, whereas in another there might only be reference to arbitration. As stated before, the situation is even more complex in that a definition of, say, conciliation in one document might be different from a definition in another, and the applicable rules/procedures may vary from the virtually non-existent to being substantially detailed.

During the 1990s, the Government and the construction industry jointly commissioned Sir Michael Latham to produce a report on the construction industry. One of the consequences of that report was that construction adjudication provisions were included within an Act then going through Parliament. On 1 May 1998, Part II of that Act, the Housing Grants, Construction and Regeneration Act 1996 (hereafter the Construction Act) came into force. This, *inter alia*, granted any Party to a construction contract (which the Construction Act defined) the right, if it wanted to exercise it at any time, to refer any dispute arising under a contract to fast-track adjudication in accordance with eight minimum requirements set out in the Act. The Construction Act and the eight requirements are discussed in detail in Chapter 2.

The major organisations in the construction industry revised their own adjudication conditions to comply with the Act. They have often enhanced the Act's eight minimum requirements to better reflect what they preferred to have in their conditions of contract. The result is that for the majority of construction contracts, the adjudication provisions of the relevant conditions of contract will apply. Three of the main contractual adjudication provisions are discussed in Chapters 3 to 5. These are the ICE Adjudication Procedure (1997), the JCT Standard Form 1998 Edition Clause 41A Adjudication Procedure and the GC/Works/1 (1998) Adjudication Procedure respectively. In addition, the stand-alone CIC Model Adjudication Procedure (Second Edition) is included in Chapter 6. There are other procedures, such as those of the Technology Court Solicitors Association (the TeCSA Adjudication Rules) and the Centre For Dispute Resolution (the CEDR Rules), but as these currently are not thought to have been so widely utilised, they are not considered in any detail within this text.

Where there are no contractual adjudication procedures, or the latter are considered not to comply with one or more of the eight minimum requirements given in the Construction Act, then the procedures stated in the Scheme for Construction Contracts (England and Wales) Regulations 1998 (hereafter the Scheme) become available in part or whole instead, if a Referring Party prefers to use them (such issues being discussed further in the main text). The Scheme is discussed in detail in Chapter 7. In fact, following substantial consultation, the Scheme (and perhaps the Construction Act) may be the subject of some legislative amendments. When that may happen, or what any final changes might be, is not currently known. However draft proposed amendments relating to England are contained within the Scheme for Construction Contracts (Amendment) (England) Regulations 2001 (hereafter the Draft Scheme Amendments). These are introduced within the main body of the text as indicative of what might happen where considered appropriate, as well as being discussed in more detail specifically in Chapter 8.

The Construction Umbrella Bodies Adjudication Task Group, (hereafter called the Umbrella Group), has produced a draft document, dated 2 May 2002, entitled *Adjudication under the Scheme – Guidance to Adjudicators* (hereafter referred to as the Draft Umbrella Group Guidance to Adjudicators using the Scheme). This document provides guidance on aspects of the statutory Scheme adjudication procedures only, although some of that guidance will also be equally relevant to other contractual procedures. It is based on the Draft Scheme Amendments and what is understood to be current thinking on those Amendments. Relevant extracts of this document are also introduced within the main body of the text where considered appropriate.

The main text of this book is divided into three Parts:

Part I provides details of the legal requirements for construction adjudication, describing the foregoing Construction Act, contractual adjudication procedures and the Scheme. The approach taken in the text is that the four contractual Adjudication Procedures and the Scheme are intended to each be as self-contained as possible, even if this means there is some duplication of wording between them. This generally enables a reader only principally interested in, say, the GC/Works Adjudication Procedure, Chapter 5, to proceed directly from Chapter 2 on the Construction Act to Chapter 5, bypassing the intervening Chapters 3 and 4 on the ICE and JCT Procedures, without missing any relevant commentary contained in those bypassed chapters.

Part II deals with the procedural conduct of an adjudication. In Chapters 9 to 13 it deals with key issues from the occurrence of a dispute through the appointment of an Adjudicator to the production of an Adjudicator's decision.

Part III is concerned with some more general issues that do not readily fit into Parts I or II, including insurance aspects. These are covered in Chapters 14 to 16.

There are several Appendices. Appendices 1 to 5 provide some examples of possible adjudication letters, generally not to be used strictly, but as aide-mémoire checklists.

This is followed by a list of non-adjudication legal cases referred to in the text, then a list of adjudication cases up to the time of writing. There are selective summaries of the details and decisions reached in those adjudication cases.

A short Bibliography of reference material is provided at the end of the text, which it is hoped will form a useful basis for further reading about adjudication-related topics.

Part I

Legislation and formal adjudication procedures

Chapter 2

The Construction Act

General

Part II of the Housing Grants, Construction and Regeneration Act 1996 (herein generally referred to as the Construction Act) came into force on 1 May 1998. It contains, *inter alia*, adjudication provisions which apply to the majority of *construction contracts* which have to be *in writing* and must be for the carrying out of *construction operations*. The following gives some of the details of the Construction Act (in italics), with additional commentary.

Section 104 Construction contracts

(1) *In this Part* [of the Act] *a 'construction contract' means an agreement with a person for any of the following –*

(a) *the carrying out of construction operations;*

(b) *arranging for the carrying out of construction operations by others, whether under sub-contract to him or otherwise;*

(c) *providing his own labour, or the labour of others, for the carrying out of construction operations.*

(2) *References in this Part* [of the Act] *to a construction contract include an agreement –*

(a) *to do architectural, design, or surveying work, or*

(b) *to provide advice on building, engineering, interior or exterior decoration or on the laying-out of landscape,*

in relation to construction operations.

(3) *References in this Part* [of the Act] *to a construction contract do not include a contract of employment (within the meaning of the Employment Rights Act 1996).*

(4) *The Secretary of State may by order add to, amend or repeal any of the provisions of subsections (1), (2) or (3) as to the agreements*

which are construction contracts for the purposes of this Part or are to be taken or not to be taken as included in references to such contracts.

(5) *Where an agreement relates to construction operations and other matters, this Part* [II of the Construction Act] *applies to it only so far as it relates to construction operations.*

An agreement relates to construction operations so far as it makes provision of any kind within subsection [104] *(1) or* [104] *(2).*

In summary, therefore, the Construction Act applies to contracts which relate, in whole or in part, to the carrying out or providing of labour, or arranging to be carried out by others or provision of labour by others, of 'construction operations'. These are defined further in Section 105. A construction contract includes professional services agreements and agreements to provide advice on internal and external building, engineering, decoration and landscape layouts in connection with construction operations.

The issue of whether legal advice agreements on construction claims are included has been clarified by the courts (*Fence Gate* v *Knowles* [2001]). It was stated (in respect of an adjudication concerning the payment of fees arising from an arbitration) that providing litigation support at an arbitration is not the same as providing advice on a building or engineering claim. Indeed, disputes in relation to payment of fees properly payable for services rendered as a witness of fact or assisting at an arbitration or litigation are not disputes in relation to construction operations – they are disputes in relation to litigation support work and arise under a contract to provide litigation support services. Contracts of employment are, however, excluded by a separate Exclusion Order, which may itself be amended later by the Secretary of State.

It will be appreciated that situations may occur where a dispute might arise in which a decision is required under several headings, only some of which might be related to carrying out construction operations in a written construction contract. Subsection 104 (5) is usually interpreted to mean that the Construction Act applies to those parts that do comply and not to those that do not. This is considered in more detail under the heading of 'Mixed Construction Act-compliant and non-compliant disputes' in Chapter 14.

Section 105 Construction operations

Section 105 (1) Types of work which are construction operations

In this Part [of the Act] *'construction operations' means, subject as follows, operations of any of the following descriptions –*

(a) *construction, alteration, repair, maintenance, extension, demolition or dismantling of buildings, or structures forming, or to form, part of the land (whether permanent or not);*

(b) *construction, alteration, repair, maintenance, extension, demolition or dismantling of any works forming, or to form, part of the land, including (without prejudice to the foregoing) walls, roadworks, power-lines, telecommunication apparatus, aircraft runways, docks and harbours, railways, inland waterways, pipe-lines, reservoirs, water-mains, wells, sewers, industrial plant and installations for purposes of land drainage, coast protection or defence;*

(c) *installation in any building or structure of fittings forming part of the land, including (without prejudice to the foregoing) systems of heating, lighting, air-conditioning, ventilation, power supply, drainage, sanitation, water supply or fire protection, or security or communications systems;*

(d) *external or internal cleaning of buildings and structures, so far as carried out in the course of their construction, alteration, repair, extension or restoration;*

(e) *operations which form an integral part of, or are preparatory to, or are for rendering complete, such operations as are previously described in this subsection, including site clearance, earthmoving, excavation, tunnelling and boring, laying of foundations, erection, maintenance or dismantling of scaffolding, site restoration, landscaping and the provision of roadways and other access works;*

(f) *painting or decorating the internal or external surfaces of any building or structure.*

Construction operations thus include construction, alteration, repair, maintenance, extension, demolition or dismantling of buildings or structures which are, or will be, part of the land, including walls, roadworks, power lines, airport runways, telecommunication apparatus, railways, docks and harbours, inland waterways, pipelines, reservoirs, land drainage installations and machinery, coastal protection works, sewers, water mains and wells, site clearance, earthmoving, excavation, tunnelling and boring, foundations, scaffolding, site restoration and landscaping, access works, building services, external and internal cleaning during construction operations, painting and decorating.

Buildings and/or structures and/or works 'forming, or to form, part of the land'

The phrase 'forming part of the land' has been the subject of considerable legal debate.

Cheshire and Burn's *Modern Law of Real Property* (Fifteenth Edition) states, *inter alia*:

> *'This question whether a chattel has been so fixed to land as to become part of it ... is a question of law for the judge.'*
>
> *'The general rule is that a chattel is not deemed to be a fixture unless it is actually fastened to or connected with the land or building.'*

'If a superstructure can be removed without losing its identity, it will not generally be regarded as a fixture. Examples are a Dutch barn, consisting of a roof resting upon wooden uprights, the uprights being made to lie upon brick columns let into the ground. ... The case is the same if the posts that support the roof of a corrugated building are not embedded in the concrete floor, but are held in position by iron strips fixed into the floor.'

On the basis of the Dutch barn example it could therefore be argued that simple non-moment resisting holding-down bolts for locating purposes are the same as iron strips, and so this could be used to exclude a range of precast concrete or steel structures from the Act. It is something to be remembered, but probably of limited application, particularly bearing in mind some of the practical adjudication implications discussed earlier.

In a court case (*Potton Developments* v *Thompson* [1998]), nine self-contained bedroom units which looked like fixed buildings at a motel inn could, in fact, be removed intact from their bolted base slabs without being dismantled, hence being agreed by the court not to be fixed to the land.

A critical implication in this respect, to be borne in mind perhaps by an Adjudicator when considering a potential decision, is that something that is fixed to the land (or incorporated into a building) is, under land law, deemed to have become the property of the land or building owner, even if it has not yet been paid for. This is not withstanding any contractual clause that states that title only passes to the owner on payment being made! The supplier cannot therefore then remove the unpaid-for item without permission from the new owner!

These examples will indicate that this area may need to be considered with a bit more care than was perhaps realised. It could, for instance, be particularly relevant to sub-contract disputes relating to whether additions to sites and property (e.g. temporary or hired-in buildings) should be classed as fittings.

One relatively straightforward adjudication decision relating to this subject has been made by the courts (*Staveley* v *Odebrecht* [2001]). It was decided that under the Construction Act, structures on the seabed below the low water mark of England, Scotland and Wales (e.g. offshore) do not form part of the land.

In another (*Gibson* v *Makro* [2001]) the courts dealt with the phrase 'fittings forming part of the land' in subsection 105 (1)(c). It was decided that shop fitting in this situation did not amount to a construction operation and that the shop fittings supplied were not fixtures.

Section 105 (2) Types of work excluded from construction operations

The following operations are not construction operations ... –

(a) *drilling for, or extraction of, oil or natural gas;*

(b) *extraction (whether by underground or surface working) of minerals; tunnelling or boring, or construction of underground works, for this purpose;*

> (c) *assembly, installation or demolition of plant or machinery, or erection or demolition of steelwork for the purposes of supporting or providing access to plant or machinery, on a site where the primary activity is –*
>
> > (i) *nuclear processing, power generation, or water or effluent treatment, or*
> >
> > (ii) *the production, transmission, processing or bulk storage (other than warehousing) of chemicals, pharmaceuticals, oil, gas, steel or food and drink;*
>
> (d) *manufacture or delivery to site of –*
>
> > (i) *building or engineering components or equipment,*
> >
> > (ii) *materials, plant or machinery, or*
> >
> > (iii) *components for systems of heating, lighting, airconditioning, ventilation, power supply, drainage, sanitation, water supply or fire protection, or for security or communications systems,*
>
> *except under a contract which also provides for their installation;*
>
> (e) *the making, installation and repair of artistic works, being sculptures, murals and other works which are wholly artistic in nature.*

In practice, it has been found the foregoing list and wording, some of which was the result of intense lobbying, is not as clear as perhaps the drafters intended and that there are peripheral areas that need clarification. For example, by way of further explanation, the Government has stated that the exclusion of process industries from the Construction Act was intended to be related to 'only work on the machinery and plant that is highly specific to the process industry, together with work on steelwork that is so intimately associated with that plant and machinery that it could not possibly be reasonably be considered apart'. The steelwork mentioned in the exclusion 'is only that which relates to support and access ... all normal construction activities on a processing engineering site will be subject to the provisions of the Bill'. There may, therefore, be some building sites where some works are covered by the Construction Act and others not.

The courts have also considered several cases where clarification was required. For example, they decided (*Palmers* v *ABB Power* [1999]) that the word 'plant' includes the pipework linking various pieces of equipment excluded by the Act and so work directly associated with the pipework is also not a construction operation.

On the other hand, the courts also stated (*Palmers* v *ABB Power* [1999]) that scaffolding required for the erection of a steam boiler on a power generation project is a construction operation, even though the assembly of the boiler itself, since the site's primary activity was power generation, was itself excluded by the Construction Act.

The courts (*Nottingham* v *Powerminster* [2000]) stated that the maintenance and repair of domestic gas appliances is a 'construction operation' and this also applies to annual service contracts and call-out repair and breakdown services.

In another case (*ABB Power* v *Norwest Holst* [2000]), the courts decided that the supply and installation of insulation was not a 'construction operation'. The latter was based on the argument that if the installation work (which would otherwise have been an exception under the last line of Section 105 (2)(d)) was related to excluded operations under Section 105 (2)(a), (b) and (c) (which it was), then the installation work was also excluded.

A later case (*ABB Zantingh* v *Zendal* [2000]) also involved installation. This time the supply, installation and testing of electrical equipment was for standby generators in an enclosed area within the site of a printing works. Notwithstanding that the enclosed area constitutes a small power generation site itself and as such might have fallen within the exclusion of Section 105 (2), the courts decided that the primary activity of the site as a whole was printing. As a result, the electrical equipment contract was deemed to be a valid 'construction operation' under the Construction Act.

In a similar decision, the Scottish courts (*Mitsui Babcock, Petitioners* [2000]) looked at a decision by a potential adjudication that a dispute lay outside the Construction Act. In this case, two sites for assembling boiler plants were in separately contained areas but situated on vacant land within a petrochemical complex, their ultimate use being to further the primary activity of the processing of chemicals and oil at the complex. It was considered that the exemption in Section 105 (2)(d) was directed to the primary activity on a site and that in this case, the potential Adjudicator was correct to decide it had no jurisdiction.

From these two cases it can be concluded that the courts do not recognise one smaller site lying inside another for the purposes of the Construction Act. What they are concerned with is the primary activity of the main site as a whole.

The existence of a 'construction contract' is an essential prerequisite for an adjudication under the Construction Act. It goes directly to the validity of the jurisdiction of the Adjudicator who might become involved, i.e. if there is no 'construction contract' there is no adjudication under the Act nor does an Adjudicator have any related jurisdiction. (See also Chapter 10 dealing with Jurisdiction.) Whether or not a 'construction contract' exists has been described as a matter for the courts. That is not to say that an Adjudicator cannot examine such an issue, but it means in practice that any decision which the Adjudicator might come to on such an issue may not have the temporarily binding effect which the Adjudicator's decision upon an issue of fact would otherwise have by statute (e.g. the decision might not be enforced by the courts). Thus, where an issue about the existence or otherwise of a 'construction contract' is raised then the Parties have a choice. The Parties could seek from the courts at the outset (the Technology and Construction Court has special procedures in this respect) a declaration that a 'construction contract' is/is not in existence. In practice, the Parties will often agree to delay any adjudication until that issue has been resolved. In the absence of any such agreement, however, a successful injunction to forestall any adjudication proceeding further, while possible, is unlikely. It may, therefore, sometimes happen in practice that the adjudication has to proceed with that issue unresolved. In these circumstances, it is vital that the Party concerned maintains its objection throughout the adjudication. Provided that is done, then thereafter, either of their own volition or in defence to enforcement proceedings, that Party could ask the Court for a declaration ruling upon the existence or otherwise of a 'construction contract'. In the event that the Court rules that there was no 'construction contract' in existence, then clearly any decision by the Adjudicator will not be enforced and that Party will, hopefully, recover its costs. Where there is lack of clarity, not assisted by current case law on the subject, it is suggested that the Parties should stay the appointment of an Adjudicator, or at least not proceed further with an adjudication, until they have asked for a ruling by the courts.

Section 105 (3) Amendments by Secretary of State

The Secretary of State may by order add to, amend or repeal any of the provisions of subsection [105] *(1) or (2) as to the operations and work to be treated as construction operations for the purposes of this Part.*

Section 106 Exclusions from construction contracts

Sections 106 (1) and (2) Residential occupier contracts excluded

(1) This Part [of the Act] *does not apply –*

(a) to a construction contract with a residential occupier (see below), or

(b) to any other description of construction contract excluded from the operation of this Part by order of the Secretary of State.

(2) A construction contract with a residential occupier means a construction contract which principally relates to operations on a dwelling which one of the parties to the contract occupies, or intends to occupy, as his residence. In this subsection 'dwelling' means a dwelling-house or a flat; and for this purpose –

'dwelling-house' does not include a building containing a flat; and

'flat' means separate and self-contained premises constructed or adapted for use for residential purposes and forming part of a building from some other part of which the premises are divided horizontally.

In other words, construction contracts with a residential occupier, i.e. where one of the Parties to the contract resides, or intends to reside in the building or flat which is the principal purpose of the contract, are excluded. It is clearly the present intention of Parliament to exclude those construction contracts which are with residential occupiers. There has been a limited interpretation of this provision so far, but it seems clear that normally a developer will not be a 'residential occupier' and therefore any contract between a developer and a contractor will be subject to adjudication. Where a person intends to develop part of a property for commercial purposes and part of a property for its residential occupation then it may be that adjudication will apply to the former but not to the latter. Where a person is contracting solely to create a 'dwelling' which it intends to occupy as its residence, then statutory adjudication will not be available. Again, an injunction to forestall an adjudication from proceeding further may or may not be available in these circumstances but a declaration stating that the construction contract in question is or is not with a residential occupier could be sought from the court either before or after an adjudication.

A court case (*Thomas* v *J & B Developments* [2000]) related to a mixed development comprising property being developed for both personal, residential and commercial purposes. The courts found that as the majority of the work, around 65% in cost terms, was related to the part of the development that would become owner-occupied as a dwelling, it could come under the Construction Act. This applied even if the original property being converted was not a dwelling before its conversion, or if the future owner-occupier was not in residence during the construction work.

Sections 106 (3) and (4) Specific exclusions by Secretary of State

The Secretary of State may by order amend subsection [106] *(2).*

Specific exclusions from construction contracts made by the Secretary of State presently include Section 38 and Section 278 Agreements under the Highways Act; Section 106 under the Planning Act; Section 104 under the Water Industry Act; externally financed developments under Section I of the National Health Service; private finance initiative (PFI) contracts and finance agreements. The latter means agreements in which the principle obligations include an undertaking by a person to be responsible as surety for a debt or default of another person, including a fidelity bond, advance payment bond, retention bond or preference bond.

PFI concession agreements

PFI concession agreements are excluded, but not any construction contract itself or any further downstream subsidiary construction contracts.

Partnering agreements

Although partnering agreements are often stated to be non-binding, they may fall under the heading of construction contracts. This seems possible where the terms of an agreement are contained within a construction contract itself. An Adjudicator might be asked to decide on the matters relating to those terms. Even if the partnering agreement is outside the main construction contract, if it contains terms relating to the management of the construction contract this may bring it under the 'carrying out' or 'arranging to be carried out' wording of this Section.

Bonds

Performance bonds are not construction contracts and so are excluded from the Construction Act, i.e. there need be no provision for adjudication in the wording of the bond nor will there be an option to use the Construction Act adjudication provisions.

However, it will be appreciated that bonds relating to costs arising as a consequence of contractual default may be called upon as a result of an adjudication decision. Bond wordings should, therefore, be reviewed to determine whether the bond is to be payable immediately to the full extent of an adjudication award or whether there may be scope for protracted negotiations regarding monetary deductions arising out of matters which were not put before an Adjudicator.

There is, however, no reason why contractual adjudication provisions cannot be included within bonds as they might prove a useful way for the signatories of the bond to resolve between them any dispute, and the relevant quantum.

Insurances

Although not specifically excluded, it appears safe to consider construction insurance policies as somewhat similar to bonds (in fact, they could both be issued by the same insurance company) which do not fall within the scope of the Construction Act.

Collateral warranties, novations

Collateral warranties and novations would not normally appear to be construction contracts under the Construction Act, except where they are signed after the Construction Act came into force on 1 May 1998 and relate to outstanding liabilities and/or obligations which are ongoing construction operations (e.g. *Yarm* v *Costain* [2001]). For example, a collateral warranty signed by an architect or engineer after professional services or advice have been completed would not appear to be a Section 104 construction contact. That does not, of course, prevent the inclusion within a collateral warranty or novation of adjudication procedures which comply with the Construction Act, or which are similar or identical to the Scheme or to other adjudication procedures.

Section 107 Contracts in writing

Section 107 A contract must be 'in writing' to be a construction contract

> (1) *The provisions of this Part* [of the Act] *apply only where the construction contract is in writing, and any other agreement between the parties as to any matter is effective for the purposes of this Part only if in writing.*

The expressions 'agreement', 'agree' and 'agreed' shall be construed accordingly.

(2) *There is an agreement in writing –*

(a) *if the agreement is made in writing (whether or not it is signed by the parties),*

(b) *if the agreement is made by exchange of communications in writing, or*

(c) *if the agreement is evidenced in writing.*

(3) *Where parties agree otherwise than in writing by reference to terms which are in writing, they make an agreement in writing.*

(4) *An agreement is evidenced in writing if an agreement made otherwise than in writing is recorded by one of the parties, or by a third party, with the authority of the parties to the agreement.*

(5) *An exchange of written submissions in adjudication proceedings, or in arbitral or legal proceedings in which the existence of an agreement otherwise than in writing is alleged by one party against another party and not denied by the other party in his response constitutes as between those parties an agreement in writing to the effect alleged.*

(6) *References in this Part to anything being written or in writing include its being recorded by any means.*

It has been postulated (*The Law of Restitution* by Goff and Jones) that entitlement to payment under an oral instruction arises by virtue of an 'equitable doctrine of restitution', not contract. Thus an oral instruction is not a construction contract and so the Act and Scheme should not apply.

It is clear, however, that Section 107 is very flexible. The Section states that a contract should be considered to be in writing in a range of circumstances, including whether it was signed or not, was by an exchange of letters or was evidenced in writing. In practice, the Section can make many construction agreements become contracts in writing.

The courts have confirmed in one case (*A & D Maintenance* v *Pagehurst* [1999]) that although there was no written contract, both Parties were proceeding as if there was one and neither Party denied a contract was in place, therefore there was an agreement in writing.

However, in another case (*Grovedeck* v *Capital Demolition* [2000]), although it was agreed by the Parties there was an oral agreement, prior to the appointment of an Adjudicator one of the Parties always refuted that the agreement was intended to be a written contract under the Act. The courts agreed that in the circumstances, no construction contract in writing existed at the time of the Adjudicator's appointment.

In a further case (*Atlas Ceiling* v *Crowngate* [2000]), work commenced on the basis of a letter of intent. Both Parties agreed that various contractual issues remained to be resolved before a contract (or a sub-contract in this case) could be entered into. The courts agreed that on the basis of what was stated in the letter of intent and in subsequent correspondence, the contract was not actually entered into until the outstanding issues were later resolved.

In yet another case (*RJT Consulting Engineers* v *D M Engineering* [2001]), the courts decided that although there was initially only an oral contract, the existence of the agreement, the Parties to the agreement, the nature of the work and the agreed price were all to be readily found within the written documentation, including fee accounts, minutes of meetings and other correspondence. It was decided that there was no need for the basic terms of an agreement to be initially in writing as the extensive subsequent documentary evidence was sufficient to bring the contract within the Construction Act. However, this went to the Court of Appeal and was unanimously overturned in March 2002. Essentially, two judges agreed that in order for an Adjudicator to decide a dispute, the whole of an agreement must be written down, except for where the parts relevant to a dispute are included within the written submissions of a Referring Party and are not disputed by the Responding Party. The third judge, although agreeing with the decision of the other two, considered that only the parts of an agreement relevant to the dispute needed to be written down.

Section 108 Minimum adjudication provisions

Section 108 (1) A right to adjudication

> *A party to a construction contract has the right to refer a dispute arising under the contract for adjudication under a procedure complying with this section.*
>
> *For this purpose 'dispute' includes any difference.*

In relation to arbitration, a body of case law had developed seeking to draw a distinction in law between a 'difference' and a 'dispute'. The foregoing provision was made to avoid such arguments applying to adjudications. The courts have considered disputes (*Fastrack* v *Morrison and Imregia* [2000]). Essentially, the matter must be put by one Party to the other Party and the latter must be given an opportunity to respond. Once that has been done, almost any non-agreement will be sufficient to give rise to an adjudication and it make no difference whether that non-agreement is characterised as a 'difference' or as a 'dispute'.

Where adjudication is provided for in a contract then it is normally expressed as an entitlement (using the words 'may refer the matter for adjudication') rather than an obligation (using the words 'shall refer the matter for adjudication').

Where the contract is silent as to adjudication, or the contractual adjudication provisions are 'non-complying' but the statutory conditions (e.g.

there is a 'construction contract') set out in the Construction Act are otherwise all met, then adjudication under the statute is also not an obligation but is an entitlement (expressed as a 'right'). What that means is that although the Party satisfying the statutory conditions could call for a statutory adjudication if it wants to, it does not have to if it chooses not to do so.

Sections 108 (2) to 108 (4) The criteria for an acceptable adjudication procedure

(2) The contract shall –

> *(a) enable a party to give notice at any time of his intention to refer a dispute to adjudication;*
>
> *(b) provide a timetable with the object of securing the appointment of the Adjudicator and referral of the dispute to him within 7 days of such notice;*
>
> *(c) require the Adjudicator to reach a decision within 28 days of referral or such longer period as is agreed by the parties after the dispute has been referred;*
>
> *(d) allow the Adjudicator to extend the period of 28 days by up to 14 days, with the consent of the party by whom the dispute was referred;*
>
> *(e) impose a duty on the Adjudicator to act impartially; and*
>
> *(f) enable the Adjudicator to take the initiative in ascertaining the facts and the law.*

(3) The contract shall provide that the decision of the Adjudicator is binding until the dispute is finally determined by legal proceedings, by arbitration (if the contract provides for arbitration or the parties otherwise agree to arbitration) or by agreement.

The parties may agree to accept the decision of the Adjudicator as finally determining the dispute. [It will be noted that this sentence is not usually considered to be an essential requirement for inclusion in adjudication provisions (but it could be argued otherwise).]

(4) The contract shall also provide that the Adjudicator is not liable for anything done or omitted in the discharge or purported discharge of his functions as Adjudicator unless the act or omission is in bad faith, and that any employee or agent of the Adjudicator is similarly protected from liability.

In summary, the eight minimum requirements are:

1. A Party must be able to give notice at any time of its intention to refer a dispute to adjudication.

2. A timetable must be provided with the object of securing the appointment of an Adjudicator and the referral of the dispute to that Adjudicator within 7 days of the aforementioned notice.
3. The Adjudicator is required to reach a decision within 28 days of the referral or such period beyond the 28 days as is agreed by the Parties after the dispute has been referred.
4. The 28-day period can be extended by the Adjudicator by up to 14 days with the consent of the Party referring the dispute.
5. The Adjudicator shall act impartially.
6. The Adjudicator may take the initiative in ascertaining the facts and the law.
7. The decision of the Adjudicator is binding until the dispute is finally determined by legal proceedings, by arbitration (if the contract provides for arbitration or the Parties otherwise agree to arbitration) or by agreement.
8. Neither the Adjudicator, nor its employees or agents, are to be liable for anything done or omitted whilst acting as an Adjudicator, unless the act or omission is in bad faith.

Section 108 (5) Scheme to apply if a minimum requirement is not met

> *If the contract does not comply with the requirements of subsections* [108] *(1) to (4), the adjudication provisions of the Scheme for Construction Contracts apply.*

This means that if (a) any one of the eight criteria is not met by the Parties' own contractual adjudication provisions, and then if (b) the option under Section 108 (1) to use Construction Act-compliant procedures is taken up by a Referring Party, the Scheme would automatically apply instead. That would clearly also be the case if there were no adjudication provisions in the contract at all. The Scheme is considered in detail in Chapter 7.

There are differing views on whether the whole of the Scheme should be substituted for a procedure that is not Construction Act-compliant, or just the part of the Scheme that relates to the non-compliant part of the procedure. It will be noted that the non-compliance may be of a minor nature or have no impact on the dispute being referred. This issue is discussed later in this chapter where various aspects of Section 108 are considered in more detail.

It is worth noting that the procedure under which an adjudication is to run is a decision for the Parties. Thus all Parties can agree to run an adjudication under a contractual adjudication procedure, even though it was, strictly speaking, 'non-compliant' with the Construction Act. It is only where one of the Parties stands upon its entitlement to a Construction Act-compliant adjudication procedure that formal decisions relating to the previous paragraph may need to be taken.

Section 108 (6) Application to England, Scotland and Wales

This Section provides some advice to the 'minister making the scheme' for England, Wales and Scotland. It need not be considered further in the context of the present text.

Detailed consideration of some aspects of Section 108

Section 108 (1) states that any Party to a construction contract can refer a *dispute* arising under the contract for adjudication under a procedure complying with the Act

An important issue here is what constitutes a 'dispute'? Essentially an Adjudicator has jurisdiction over disputes, not claims. It requires a claim and a rejection (*Monmouth CC* v *Costelloe & Kemple* (1965)), or at least a reasonable maximum time by which to respond, after which a claimant can then proceed with the dispute. The courts specifically considered this issue in several adjudication cases (*Fastrack* v *Morrison and Imregia* [2000] and *Nuttall* v *Carter* [2002]). They have stated *inter alia* that a dispute arises once it has been brought to the attention of the opposing party and that party has had the opportunity of considering and admitting, modifying or rejecting it. A rejection could be considered to have arisen where a Party refused to answer a claim.

Disputes are considered in more detail in Chapter 14 Miscellaneous Issues.

Section 108 (1) Any Party to a construction contract can refer a dispute arising *under the contract* for adjudication under a procedure complying with the Act

It is necessary to understand the significance, if any, of the term 'arising under' the contract, compared with, say, 'arising out of' or 'arising in connection with' the contract. Such terms have been subject to guidance by the courts, albeit generally in arbitration-related cases.

It has been stated (*Fillite* v *Aqua Lift* [1986]) that disputes 'arising under a contract' relate to 'obligations created or incorporated into' a contract. It was further held (*Produce Brokers* v *Olympic Oil and Coke* [1916]) that strictly speaking disputes do not arise 'out of' a contract but out of the conflicting views taken by the Parties to a dispute. In other words, such disputes arise 'in connection with' or 'in relation to' the contract.

Notwithstanding the foregoing, there is a general view that 'arising out of' is wider than 'arising under' but this is not always so, and it depends on the facts of the case concerned (*Antonis P Lemos* [1985]).

The situation is, however, clearer on what is referred to as 'contractual negligence', albeit case law is generally arbitration-related.

The courts (*Abdullah Fahem* v *Mareb Yemen Insurance and Tomen (UK)* [1977]) have used the term 'contractual negligence' where contractual and non-contractual liabilities are linked when a lack of duty of care and skills has led to a deficiency in the performance of a contract. Further, the courts (*Al-Naimi* v *Islamic Press Agency* [1998]) decided that there was no effective difference between disputes arising 'under' and 'in connection with' a contract when considering whether to stay an action to arbitration.

In another case, it has been decided that 'when the contractual and tortious disputes are closely knitted together on the facts, the agreement to arbitrate one may properly be construed as covering the other' (*Empressa Exportadora de Azucar* v *Industria Azucarera Nacional SA* [1983]).

Mustill & Boyd (Second Edition) has also considered this matter. It states that: 'Most instances of claims in tort submitted to arbitration relate to 'contractual negligence', i.e. the breach of a duty of care arising from a contract. Most of the more common forms of arbitration clause are sufficiently wide to give the Arbitrator jurisdiction over such claims'.

It is therefore considered that, notwithstanding the actual wording (e.g. 'arising under' as opposed to 'arising in connection with'), an Adjudicator will also be able in practice to decide a dispute involving 'contractual negligence' (i.e. a breach of an implied duty arising from a contract) as well as a breach of contract. Indeed, contracts commonly include tortious duty of care wordings in any event. For example, a professional services contract will invariably refer to an obligation of the provider of services to exercise reasonable skill and care.

Any contract will have its own definition determining the scope of its contractual adjudication procedures. The precise ambit of the term 'arising under the contract' as it relates to statutory adjudication has not yet been fully tested by the courts. However, the courts (*Northern Developments* v *Nichol* [2000]) have already decided that the phrase is wide enough to cover damages for repudiation (i.e. unilaterally giving up the contract) and the indications are that this definition is likely to receive a wide interpretation.

Another interesting aspect arises from an arbitration case (*Harbour Assurance* v *Kansa* [1993]). It was decided that words in a contract allowing an Arbitrator to deal with all disputes arising 'out of or in connection with' a contract permitted the Arbitrator to deal with matters that concerned the actual existence of that contract (i.e. whether it was a valid contract or not), whereas if the wording referred only to disputes arising 'under' a contract, the Arbitrator was limited, inter alia, to deciding matters of interpretation in what was a valid contract.

Section 108 (2)(a) A Party must be able to give notice of an intention to refer a dispute to adjudication *at any time*

It is important to realise that the notice to refer to adjudication must relate to the dispute that has arisen, without any further issues tacked on that are not disputed, e.g. that have not been presented to the other side with an opportunity for that side to consider, modify or reject what is being claimed.

Certain adjudication procedures effectively define a dispute as not arising until other actions have taken place first. (See Chapter 3 on the ICE Adjudication Procedure, Chapter 14, and also the discussion on disputes under Section 108 (1) of this chapter.) Whether such procedures, which are described by some as 'delaying mechanisms', comply with this Section's 'at any time' requirement will be a matter for the courts, but present indications are that they would be considered to be non-compliant.

The phrase 'at any time' has been the subject of several court cases (*A & D Maintenance* v *Pagehurst* [1999], *Straume* v *Bradlor* [1999] and *Herschel Engineering* v *Breen* [2000]). The courts have confirmed that an adjudication can be commenced after a contract has been determined, or if arbitration or litigation on the same issue is ongoing, but that adjudication cannot be brought against a company in administration without leave of the court first.

There is, of course, a limiting start date of 1 May 1998, which is when the Construction Act came into operation. It is a necessity for the contract concerned to be 'entered into' on 1 May 1998, or thereafter, for the Act to comply. The courts (*Atlas Ceiling* v *Crowngate* [2000]) considered the case of a letter of intent signed before 1 May 1998, followed by a formal contract being signed later than 1 May 1998. That contract was to have retrospective effect so that the earlier works were subject to the same conditions as the later works. It was decided that although the Parties' rights and obligations in respect of the earlier work was retrospective, this did not make the date the contract was 'entered into' retrospective for the purposes of the Construction Act.

Conversely, there is no definition of when the Construction Act is to cease to apply. It is therefore likely that Practical Completion or a Certificate of Final Completion will be of no legal effect on this issue, and that the right to adjudicate at any time will pass right through the defects, retention and liability period, and will be ended only by the legal limitation periods applicable to other similar claims.

Section 108 (2)(b) A timetable *shall* be provided *with the object of* securing the appointment of an Adjudicator and the referral of the dispute to that Adjudicator within 7 days of the aforementioned notice

The Construction Act requires that an adjudication procedure shall contain a timetable that enables a referral to be made within 7 days. Whilst some commentators believe that the use of the word 'shall' imposes a mandatory requirement for appointment and referral within 7 days, others contend that the meaning is just that it should be possible to appoint and refer within that time scale. The latter seems to be usually followed in practice, but this has not yet been tested by the courts. This is similarly discussed under the heading of 'Urgency of appointment' in Chapter 9.

Section 108 (2)(c) Decision to be reached within 28 days of referral, or longer if agreed by the Parties

What is meant by 'reached' has been the subject of some debate. Does it mean 'reached and provided without any delay to the Parties' or does it mean 'reached and available to the Parties once any pre-agreed obligations, such as the payment of any outstanding Adjudicator's fees, have been met'. This is discussed further under 'Retention of decision until fees are paid' in Chapter 13, the Decision.

Section 108 (2)(d) The 28-day timetable period can be extended by the Adjudicator by 14 days if the Referring Party agrees

It is not clear why only the Referring Party can agree to a 14-day extension. There might be occasions when, for example, a Referring Party is not complying with Adjudicator's instructions, is supplying information that to be fair needs a adequate opportunity to reply to be given by the Responding Party, or a cross-claim (see later) by the Responding Party can only be considered by the Adjudicator if it has an extension that the Referring Party is not interested in providing.

It is interesting to contemplate whether, if a contractual adjudication was amended to also allow a Responding Party an identical or similar right, this would be considered a legitmate enchancement to the minimum requirements of the Construction Act. On the face of it, this would appear to be the case.

Section 108 (2)(e) An Adjudicator *shall act impartially*

Impartiality means not to act partially or biased. It will be noted that there is no express reference to 'independence' as in the European Convention on Human Rights, or to similar wording in some other conditions of contract. This issue is dealt with in substantially more detail in Chapter 11 in the section dealing with the Rules of Natural Justice.

Section 108 (2)(f) An Adjudicator *may* take the initiative in ascertaining the facts and the law in order to reach a decision

Under this Section, an Adjudicator has the authority to do a wide range of things (as long as they are done impartially) in order to ascertain the relevant facts and law. This authority is usually listed in adjudication procedures. For example, an Adjudicator can request documents and written statements, meet and question specified representatives of the Parties, make site visits with or without the Parties, arrange for additional tests to be carried out, appoint advisors, produce an adjudication timetable, set deadlines and issue instructions.

Section 108 (3) The decision of the Adjudicator is binding until the dispute is finally determined by legal proceedings, by arbitration (if the contract provides for arbitration or the Parties otherwise agree to arbitration) or by agreement. The Parties may agree to accept an Adjudicator's decision as a final determination

It will be noted that it is not the Adjudicator's decision that might be referred to litigation, arbitration or agreement between the Parties for final determination, but the dispute that gave rise to the adjudication decision.

The Parties may agree that an adjudication decision will be the final determination of a dispute, i.e. that it will be conclusive and not be taken further, either to arbitration or litigation, or be the subject of further negotiation. In theory such an agreement need not be formal or written but it seems sensible that it should be so.

It will be noted that there is nothing expressly stated as to when such a conclusive agreement can be reached and it could be before or after an adjudication decision. In context, the statement permitting a conclusive agreement follows a statement clearly dependent on the existence of a decision and so the implication might be that a conclusive agreement should be reached after the decision is made. However, there is no obvious significant reason for such a particular restrictive requirement, and it could be argued that the intention was to point out, after the first sentence of Section 108 (3), that there was an express exception to that sentence's requirements.

It is considered that an agreement to make a decision conclusive after it has been made is less likely to happen when there is a winner and a loser, compared with one that was agreed at the contract wording stage before a dispute arose. To have the latter expressly stated within the adjudication procedures or conditions of contract would not require further agreement between the Parties. It could include pre-agreed limits, e.g. it could be agreed to apply to, say, all adjudication decisions of less than an arbitrary figure of perhaps £50,000. Alternatively, a limit might be for an amount that the Parties agree that for economic reasons is unlikely to be carried forward to arbitration or litigation, or which is below any relevant insurance excess. (The Parties should remember to ascertain their insurers' requirements, if any (see Chapter 16 on Insurance Implications), regarding the level at which they can make a conclusive agreement.)

There remains the possibility that a Referring Party might try to argue that any such prevailing (albeit freely entered into) pre-agreed procedure was not in accordance with the Construction Act, and use it as an excuse to select to use the Scheme (see Section 108 (5) following) instead. That particular matter might need to be resolved by the courts.

Notwithstanding the foregoing, a further issue that might need addressing in England and Wales is if a conclusive, final and binding agreement (whether pre or post adjudication decision) needs 'consideration' to make it valid and if so, how that should be incorporated into that agreement.

An Adjudicator's decision may also be made binding by default if the Parties carry on as if that decision was accepted. The situation can be formalised within an underlying contract, e.g. the ICE Conditions of Contract, Clause 66 (9)(b), states:

> *Where an Adjudicator has given a decision under Clause 66 (6) in respect of the particular dispute the Notice to Refer [to arbitration] must be served within three months of the giving of the adjudication decision otherwise [the latter] will be final as well as binding.*

There is a similar statement in Condition 60, Paragraph 1 of GC/Works/1 (see Chapter 5), which specifically refers to adjudication. The phrase 'the Adjudicator's decision thereon shall become unchallengeable … ' is used.

As wording such as the foregoing is part of the contract between the Parties, it is considered that such default mechanisms would constitute an agreement between the Parties complying with the first sentence of Section 108 (3).

Section 108 (4) An Adjudicator (and its employees or agents) *should not be liable for anything done or omitted in the discharge of his functions as Adjudicator*, unless the act or omission is in *bad faith*

The wording of this Section is intended to protect an Adjudicator against liability under contract, i.e. to the Parties, as a consequence of the Adjudicator performing its activities. Unlike an Arbitrator, an Adjudicator does not benefit from a statutory indemnity, and so needs an indemnity from the Parties. It will be noted that such an indemnity might be considered to be invalid where an Adjudicator acts outside its jurisdiction, i.e. outside its contract with the Parties. Although in practice it is considered unlikely that a court would find an Adjudicator liable under most circumstances (indeed if it did so the operation of the Act could be severely restricted), it is a matter that should at least be borne in mind by any Adjudicator proceeding with an adjudication after receiving an objection to its jurisdiction from one of the Parties.

An Adjudicator also has a non-contractual potential liability to third parties by whom it can be sued for negligence in tort. This can occur if, say, an Adjudicator gives a decision in relation to a specification or method of working which results in a building defect or failure that causes third party loss or injury.

One way of giving an Adjudicator protection against third parties is for the Parties to provide an effectively worded indemnity. For example, Paragraph 7.2 of the ICE Procedure, *inter alia*, requires the Parties to 'save harmless and indemnify' the Adjudicator against claims by third parties affected by the adjudication. The wording of the indemnity can thus be critical. For example, in some circumstances (the Court of Appeal in *Walters* v *Whessoe* [1960] – as mentioned with approval in the House of Lords in *Smith* v *South Wales Switch Gear* [1977]) an Adjudicator may have a liability in negligence if a contractual indemnity to be provided by Parties to the Adjudicator does not specifically include negligence. In the foregoing ICE example, it might be better if a phrase specifically mentioning negligence was included.

Finally, since the Construction Act came into force in England and Wales in May 1998, the Construction (Rights of Third Parties) Act 1999 has also appeared. The latter permits a third party to enforce the terms of a contract between the Parties when that contract expressly allows it to and there is a benefit to be gained by the third party. The Act has therefore brought with it additional potential liabilities which may need to be addressed in particular conditions of contract and/or adjudication procedures and/or Adjudicator Agreements to reinstate full protection of an Adjudicator against third party claims. However, a third party should stand in no better position than the Parties. As far as the Adjudicator is concerned, this probably means the third party can only act against the Adjudicator if the latter had acted in 'bad faith'. Essentially, it is suggested that to protect an Adjudicator, all that is needed is an express statement in the Adjudicator's Agreement to the effect that the Parties to the Agreement do not intend any of its terms to be enforceable by a third party. The Parties to the underlying contract might consider incorporating something similar to protect themselves within that contract. Indeed, to provide full protection to the Adjudicator, this might be essential in some circumstances, otherwise, notwithstanding what is in the Adjudicator's Agreement, a third party might acquire rights against that Adjudicator through the main contract!

The Parties' obligation to agree an indemnity for the Adjudicator does not extend to acts committed in 'bad faith'. 'Bad faith' may be considered to be the opposite of 'good faith'. It is 'malice in the sense of personal spite', a 'desire to injure for improper reasons' and, according to dictionaries, 'an intent to deceive'. It thus can include making a decision whilst having knowledge of an absence of power to make that decision. The foregoing is therefore frequently stated as 'bad faith or excess of power'.

Bad faith can create liability for damages and criminal prosecution and is not insurable. This, in conjunction with an indemnity provided by the Parties, does not, however, mean that there is no need for an Adjudicator to have any professional indemnity insurance cover at all. There may, for example, be some residual liability for third party claims. What it does mean is that if a professional is only providing services as an Adjudicator, then professional indemnity insurance premiums might be expected to be relatively low.

In the early days of proposed statutory adjudication, it was suggested that it might also be possible to use an 'excess of powers' liability argument to appeal an Adjudicator's decision. The basis of the argument was that powers must have been exceeded because it was impossible in the time available to consider fully and fairly all those matters necessary to establish the legal entitlement of the Parties. In fact, the courts quickly decided (*Macob* v *Morrison* [1999]) that although a tight timetable might or might not have prevented a Party having the opportunity to adequately make its case, a 'losing' Party still had to comply with an Adjudicator's decision as if it was final and binding until finally determined otherwise. The courts supported what they perceived to be the intentions of the Construction Act in that case (see also Court of Appeal decision in *Bouygues* v *Dahl-Jensen* [2000]). It is considered that, until a more appropriate case is decided, it would be prudent to anticipate a similar response by the courts to the 'excess of powers' suggestion.

In practice, bad faith is likely to be extremely difficult to prove.

Section 108 (5) If the adjudication provisions of any construction contract, including any standard conditions of contract, do not meet the minimum adjudication requirements stated in the Act, then the Scheme applies

As previously discussed, if a contract falls under the Construction Act and any one of the eight minimum requirements of the Act is not met by contractual adjudication provisions (and at the extreme, if there were no contract adjudication provisions at all) then the *option* is there to use the underlying statutory Scheme adjudication procedures.

It is clear that if there are no adjudication provisions in the contract a Party with a dispute can choose either statutory Scheme adjudication, or choose not to use adjudication. Similarly, where a contract has a provision for adjudication, but no details of what is exactly intended, a Referring Party now has the option of the Scheme provisions, or may choose not use adjudication.

Section 108 (5) is actually a subsection coming under the general heading of Section 108 of 'Right to refer disputes to adjudication'. As already discussed, Section 108 (1) provides a Referring Party to a construction contract with an option (i.e. it is not compulsory) to use a Construction Act-compliant set of adjudication procedures for the resolution of a dispute. A current common interpretation of Section 108 (5) is that 'a Party with a dispute must always use Construction Act-compliant adjudication procedures if it wants, or is obliged, to adjudicate within a contract'. The latter interpretation is reinforced by the cover sheet to the Scheme which, out of the context of following the heading of Section 108 and the optional provision of Section 108 (1), essentially simply restates Section 108 (5). However, if the Act had meant this interpretation, surely it would have been very simple to incorporate a one-sentence express statement, such as that quoted, stating just that?

It is suggested that an equally, if not more, valid interpretation is that if a contract expressly requires adjudication, a Party with a dispute in the first instance is contractually obliged to use the adjudication procedures freely agreed by the Parties in the contract, whether that contract is Construction Act-compliant or not. For example, an Employer might consider it perfectly acceptable (and perhaps potentially fairer), whether or not it is acting as a Referring or Responding Party, to have up to three months for an Adjudicator to reach a decision, and that is what is incorporated into the contract and agreed by the other Party. It is only if a Party with a dispute prefers to use procedures complying with the Construction Act (and that, *clearly stated* in Paragraph 108 (1), is for that Party to decide) that there is the further option (but only if at least one of the contract adjudication procedures is not Construction Act-compliant) for that Party to choose the Scheme instead. This thus protects a perceived 'weaker' Party which may, in retrospect, decide that for a particular dispute or disputes it wants the protection of the Scheme and not, say, Employer-produced (albeit agreed to) adjudication provisions.

The current practice of some Adjudicators certainly appears to be to imply in the whole Scheme of their own volition, but this is potentially a very significant legal step upon which, as already noted, a final pronouncement (perhaps by the Court of Appeal) has not yet been made.

An Adjudicator will usually have no express jurisdiction to select the adjudication process. It normally obtains its powers from the adjudication procedures, not the Construction Act. It cannot therefore select procedures to give it the power to select those procedures. That would clearly be a nonsense. It is therefore important for an Adjudicator to note that it would be *acting outside its jurisdiction* to impose, say, the Scheme, on the Parties if for any reason a Referring Party decides that it prefers to use any actual (or arguably) non-Construction Act-compliant procedures stated in a contract. This situation might arise, for example, where an Adjudicator does not consider that ICE matters of dissatisfaction procedures are Act-compliant (see Chapter 3 on the ICE Adjudication Procedure).

It is suggested that the only situation where an Adjudicator should become involved in deciding adjudication procedures is where the Parties themselves ask the Adjudicator to decide such a matter, e.g. is there a 'construction contract', is it 'in writing' and so on, as discussed elsewhere in this chapter. An Adjudicator might then influence whether a Referring Party decides to use its option to have a Construction Act-compliant adjudication procedure. A more direct issue that might be referred to an Adjudicator to decide is 'what is the correct (or similar wording) Adjudication Procedure for this dispute?' It would be a bold Adjudicator that did not suggest the Parties referred such an issue to the courts instead.

Also, as the adjudication is instigated by the Party with a dispute (the Referring Party) and Sections 108 (1) and (2) are specific to a Referring Party, it appears to be irrelevant that a Responding Party might prefer to use the Scheme, unless it becomes the Referring Party for the purposes, say, of a cross-claim, in which case it is an entirely different adjudication reference anyway.

To summarise the foregoing, it is considered that Paragraph 108 (1) of the Construction Act states clearly that it is effectively *only the Referring Party* that can decide on which adjudication procedures it wants to be used for its referral, i.e. not the Adjudicator unilaterally and certainly not the Responding Party.

It is noted that in at least one early adjudication case (*Macob* v *Morrison* [1999]), the Scheme has been reported as being imposed by an Adjudicator.

It will also be noted in a Scottish case (*Karl Construction* v *Sweeney* [2000]), the courts allowed just that situation, although they were 'left feeling uncomfortable'. In that case an Adjudicator decided that the payment provisions of a contract were none Construction Act-compliant and, without discussion on the issue with the Parties, proceeded to a decision using the Scheme for Scotland. The courts decided that this was a procedural mistake and not an excess of jurisdiction. The decision was supported by an Extra Decision of the Inner House of Court of Session [2002]. It is to be wondered if the English courts would reach the same conclusion in similar circumstances.

A further issue, which it is considered has not yet been finally determined (although there are some *obiter dicta* (i.e. parts of a judge's decision which are not binding as precedent) to the contrary) is whether there requires to be implied in from the underlying statutory Scheme only that which is necessary to supply the omission, or whether the underlying statutory Scheme always requires to be implied in its entirety for every defect, however minor or not relevant to the dispute being deferred.

One example of this is that an adjudication procedure might omit a requirement for an Adjudicator's employees and agents to be indemnified. Why should the Construction Act impose the whole Scheme in such circumstances? At the very most the substitution of the Scheme's equivalent requirement should be all that is required, leaving all the other Act-compliant procedures (albeit perhaps enhanced) agreed between the Parties untouched. Put another way, it is quite reasonable and admissable that, although the other minimum requirements of the Act are met, several have been considerably enhanced (and agreed between the Parties as evidenced by their signing of the contract). Is is right that such enhancements, freely entered into, should be totally disregarded by being substituted by the statutory Scheme, just because there has been a minor infringement (which might have no practical effect) of the requirements of the Act?

The adjudication position if the Construction Act does not apply

If the contract provides for adjudication anyway

Unless otherwise agreed, Parties are obliged by a contract which was freely entered into to use the contractual adjudication provisions stated in, or specifically referred to within the contract, notwithstanding that a statutory entitlement might be missing.

With some ad hoc type conditions of contract there may be specific reference to using the adjudication provisions of the Scheme. The Scheme would presumably therefore apply, even for a contract that was not a construction contract in accordance with the Construction Act.

Where the conditions of contract contain adjudication provisions complying with the Construction Act, there would, therefore, appear to be absolutely no practical reason to argue whether the contract is a 'construction contract' or not. For example, if a contract is a construction contract in accordance with the Construction Act, but a dispute is about a small part of that contract (say, involving the supply of building components that would be excluded as a 'construction contract' if it was the subject of its own contract), the overall contractual adjudication provisions (which in this case, coincidentally or otherwise, comply with the Construction Act) would still apply.

If the contract does not provide for adjudication

In these circumstances, there is no contractual entitlement to adjudication. Whatever the contract stipulates for the resolution of disputes (mediation, expert determination, arbitration or the like) should be followed, failing which resort might have to be made to litigation.

In a large project involving a number of contracts between the same Parties it will be apparent from the two foregoing situations that it is entirely possible that some of the contracts might fall outside statutory adjudication and be resolved by contractual adjudication provisions, whereas other contracts might fall within the ambit of statutory adjudication and may, if the option to use statutory adjudication procedures is selected, be resolved by such procedures. For a single project between Parties involving more than one contract, some disputes may therefore be resolved by one adjudication procedure and other disputes resolved by another adjudication or ADR procedure.

Summary

In their totality the adjudication and associated provisions of the Construction Act can appear confusing, e.g.:

- Is there a 'construction contract'?
- Is it 'in writing'?
- Do the contract provisions comply with at least the eight minimum requirements of the Act or is the option to use the Scheme available?
- What happens if some heads of claim in a dispute are not part of a 'construction contract' but some are?

In fact, the situation is far more straightforward because the majority of current standard (and non-standard) adjudication procedures have been drafted with the specific intention of complying with the minimum requirements of the Construction Act. Parties will be obliged to use those compliant contractual procedures incorporated (directly or by reference) within the conditions of contract even where the contract is not a written construction contract with construction operations as defined by the Act.

Some Parties favour the Scheme anyway. Its provisions are almost certainly better than adjudication procedures available before the Construction Act, and some sections may even now be considered better than the corresponding sections of other more recent, updated or new, procedures.

Chapter 3

ICE Adjudication Procedure

General

The Institution of Civil Engineers Conditions of Contract for Works of Civil Engineering Construction, 6th Edition (1991) [the 'ICE 6th'], Seventh Edition (1999) Measurement Version [the 'ICE 7th'], Design and Construct 2nd Edition (2001), Minor Works 3rd Edition (2001) and the associated 'Blue Form' sub-contract, each have similar adjudication requirements by incorporating by reference the ICE Adjudication Procedure (1997) [hereafter the ICE Procedure].

The ICE Procedure is drafted to comply with the Construction Act. With amendments, including provision for the naming of an Adjudicator and for a nominating body, it can be used with any non-ICE conditions of contract.

The current text considers the adjudication clauses of the ICE 6th and ICE 7th, which are identical, and refers to them both as the 'ICE Conditions of Contract' or the 'ICE Conditions' for short.

Essentially, under the ICE Conditions, a dispute arises once a matter of dissatisfaction decided by the Engineer is disputed by one of the Parties or the time for that decision has expired. (This is discussed in more detail under the heading of Section 108 (1) in Chapter 2.) If the dispute arises before a Notice to Refer to arbitration has been served, a Party can ask the other Party if the dispute can be dealt with by conciliation in accordance with the then current version of the ICE Conciliation Procedure. If so, the recommendation of the Conciliator can finally determine the dispute, i.e. it becomes 'final and conclusive' if a requisite Notice to Refer the dispute to adjudication or to arbitration is not served within one month of the conciliation recommendation. If a Notice of Adjudication is given and the adjudication proceeds, and no Notice to Refer to arbitration is received within three months of an adjudication decision being made, the latter also becomes 'final and conclusive'.

As will be discussed later, whilst arbitration (or litigation) remain the final methods of resolving disputes arising under ICE Conditions that have not been the subject of earlier final and conclusive decisions, the ICE Conditions in Clause 66 (9)(c) permit any summary enforcement of adjudication decisions to be made by the courts.

The following considers the detailed requirements. The ICE Procedure Paragraphs are quoted in full and given in italics.

Clause 66 of the ICE Conditions of Contract

The following Conditions of Contract are paraphrased and reference to the original text will be needed to see the full wording in its correct context.

Clause 66 (2) Decision of the Engineer

> *If at any time one of the Parties to the contract, (i.e. the Employer or the Contractor), is dissatisfied with any decision, opinion, instruction, direction, certificate, or valuation of the Engineer or with any other matter in connection with or arising out of the Contract or the carrying out of the Works, the matter of dissatisfaction has to be referred to and decided by the Engineer within one month.*

It will be noted that whereas the Construction Act refers to adjudication of disputes 'arising under' the contract, the Engineer (as well as a Conciliator or Arbitrator) under the ICE Conditions can deal with issues 'arising under' as well as those 'in connection with the contract'.

Clause 66 (3) Unacceptable Engineer's Decision, Notice of Dispute

> *If either Party finds an Engineer's decision unacceptable, or if a decision has not been given within one month, or if a Party has not implemented an Engineer's decision, then either Party may serve a Notice of Dispute.*
>
> *Similarly, if either Party has not given effect to an Adjudicator's decision on a dispute under Clause 66 (6), the other Party can serve a Notice of Dispute.*
>
> *The 'dispute' is the matter described in the Notice of Dispute.*

Essentially, the ICE are defining in the first part of this Clause at what point a 'claim' (say) becomes a dispute. Regarding the second part of this Clause relating to the non-observance of the Adjudicator's decision, it would now be better reworded to acknowledge that the very probable first action would be the commencement of enforcement proceedings in court.

Clause 66 (6)(a) Notice of Adjudication

> *At any time a Party may refer a dispute 'arising under' the contract to adjudication by the serving of a Notice of Adjudication to the other Party. The adjudication shall then be undertaken in accordance with the current amendment of the ICE Procedure.*

It will be noted that Paragraph 108 (1) of the Construction Act says a dispute is 'any difference' arising under the contract, whereas Clause 66 (3) states that a dispute occurs only when evidenced by a Notice of Dispute.

It has been argued that the ICE 'matter of dissatisfaction' described is a 'difference', indeed that it is a 'dispute', and that any clause that attempts to delay an entitlement to refer a matter directly to adjudication by even a month is not in accordance with the Construction Act. If correct, adjudication under the Scheme can therefore be requested instead (assuming that is what the Referring Party wants) instead of adjudication under the ICE Procedure.

The effectiveness of the ICE wording relating to what constitutes a dispute is an issue to be argued in the courts by those that want to raise it. It will be noted that the courts (*Mowlem* v *Hydra-Tight* [2000]) have already suggested that the ICE Procedure does not comply with the Construction Act in this respect. Therefore, if it is an issue decided by an Adjudicator, that decision and the jurisdiction to make it may well be reviewed by the courts.

'Disputes' are discussed in more detail under Section 108 (1) in the preceding chapter and also under the headings of 'What is a dispute' and 'Order of precedence for dispute resolution' in Chapter 14.

It will also be noted that Paragraph 5.3 of the ICE Procedure which, under Paragraph 1.1 of the ICE Procedure (see later) can take precedence over Clauses in the ICE Conditions, permits an Adjudicator to deal with issues relevant to the dispute that arise both under *and in connection with* the contract.

Clause 66 (6)(b) to (f), Clauses 66 (7) and (8) Various

Several matters that are required by the Construction Act and are also in the ICE Procedure are stated. These are the need for a 7-day Adjudicator selection timetable; to have an Adjudicator's decision within 28 days unless the Adjudicator agrees to an extension by both Parties or up to 14 days if requested by the Referring Party; the Adjudicator to act impartially and to take the initiative in ascertaining the facts and law; the Adjudicator's decision will be binding until finally determined by litigation, arbitration, or agreed between the Parties.

Clause 66 (9)(a) Arbitration (Enforcement of adjudication decisions)

There is no obligation for the enforcement of Adjudicator's decisions to be by arbitration.

The precise words actually used in the Clause might be interpreted to deliberately exclude enforcement from arbitration. Whatever the interpretation, there is no doubt that the Parties can seek enforcement from the courts instead. The courts are doing just that (*Macob* v *Morrison* [1999], *Outwing* v *Randell* [1999] etc.).

Clause 66 (9)(b) Arbitration (Final and binding adjudication decision by default)

If no Notice to Refer to arbitration is served within three months of receipt of the Adjudicator's decision, then that decision will become final and binding.

An adjudication decision can thus become final and binding by default, precluding it from any other ADR procedure including further adjudication, or arbitration or litigation (unless there has been a serious jurisdictional error or an Adjudicator has acted in bad faith).

The ICE Procedure

Paragraph 1.1 ICE Procedure and Adjudicator's Agreement. Precedence over Conditions of Contract

> *The adjudication shall be conducted in accordance with the edition of the ICE Adjudication Procedure which is current at the date of issue of a notice in writing of intention to refer a dispute to adjudication (hereinafter called the Notice of Adjudication) and the Adjudicator shall be appointed under the Adjudicator's Agreement which forms a part of this Procedure.*

i.e. the adjudication *must* be conducted in accordance with the current ICE Procedure and the Adjudicator *must* be appointed under the pro forma Adjudicator's Agreement attached to the ICE Procedure.

The Adjudicator's Agreement is considered in more detail in later Paragraph 3.4 and within Chapter 9 on Adjudicator Appointments.

> *If a conflict arises between this Procedure and the Contract then this Procedure shall prevail.*

One implication of this has already been discussed under Clause 66 (6)(a) and will be raised also within the the discussion on Paragraph 5.3.

Paragraph 1.2 Fair, rapid and inexpensive adjudication

> *The object of adjudication is to reach a fair rapid and inexpensive determination of a dispute arising under the Contract and this Procedure shall be interpreted accordingly.*

Fairness is not a requirement of the Construction Act (as discussed in Chapter 11 Procedural Fairness), e.g. the different time available to each Party to present its respective side to any dispute might not be considered to be fair. The courts on a Scheme adjudication (*Macob* v *Morrison* [1999]) where there was no stated requirement for fairness, refuted that there was any such need (although the decision in the aforementioned case as to the legal consequences of 'procedural irregularities' was subsequently reinterpreted in later cases (e.g. *Discain* v *Opecprime* [2000]). The present position might be described as being that 'serious' procedural irregularity or unfairness amounting to a breach of the Rules of Natural Justice (see Chapter 11) may result in any such decision being struck down by the courts, whereas 'minor' procedural irregularity or unfairness are something which the Parties are likely to have to live with as such decisions are usually enforced by the courts.

For contractual adjudication procedures such as the ICE, where there is an express requirement for fairness, occasional arguments about a decision being invalid because of procedural unfairness would appear to be inevitable. The extent to which judgments relating to statutory Scheme adjudications will or will not be applied to contractual adjudications, which carry an express requirement of 'fairness', remains to be seen.

The issue of disputes arising 'under' the contract, or otherwise, is discussed in more detail under the second part of the discussion on Paragraph 5.3.

Paragraph 1.3 Named individual acting impartially

The Adjudicator shall be a named individual and shall act impartially.

The words 'named individual' can cover a self-employed individual or a consultant, or a named director, or partner or individual working for a larger organisation. Indeed, the latter is no different to where a construction Arbitrator or expert is a 'named' person making personal decisions or providing personal opinions, whilst still being engaged by that organisation. This Paragraph appears far more reasonable and appropriate than, say, Paragraph 4 of the Scheme (see later).

The pre-contract award selection of an agreed named individual Adjudicator under any adjudication procedures has the advantages of speed of appointment if a dispute arises, with knowledge and pre-agreement of the terms and conditions of the appointment. There is also possibly less scope for a successful appeal by either Party on the grounds of possible bias if that was not recorded at the time of signing a contract which incorporates the agreed named individual. A disadvantage is that any one person might not be the most appropriate for a particular dispute, i.e. an Adjudicator of a particular professional discipline might be more suitable on some occasions. However, the pre-contract award selection of an agreed list of named individual Adjudicators is not precluded and has several additional practical advantages to the Parties (e.g. range of expertise and the availability of an immediate standby if one Adjudicator is not available), all at no significant additional administrative cost or time.

What is meant by to 'act impartially' is discussed in detail in Chapter 11.

Paragraph 1.4 Adjudicator may take initiative

In making a decision, the Adjudicator may take the initiative in ascertaining the facts and the law.

The adjudication shall be neither an expert determination nor an arbitration but the Adjudicator may rely on his own expert knowledge and experience.

The need in the second sentence of this Paragraph to make specific reference to 'expert determination' is not understood, nor why expert determination is expressly mentioned but not, say, conciliation or mediation. It is suggested that a definition is needed, as different ADR terms often mean different things to different persons and organisations. It is possible that this sentence is an attempt to better define the function of an Adjudicator. As stated, whilst an Adjudicator is permitted to rely on its own knowledge and experience, adjudication case law requires the Adjudicator to inform the Parties, in the interests of natural justice, when it is doing so, to give the Parties an opportunity to respond. This probably places the Adjudicator somewhere between an expert determiner (who might not have a duty to provide such disclosures) and an Arbitrator (whose own knowledge and experience will mainly be used to better understand the evidence provided by the Parties).

There will, inevitably, be adjudications that will tend more towards arbitration than expert determination, and vice versa. For example, it would be accepted generally that an expert determination comes about when two Parties agree to appoint a third party to use its knowledge and experience in order to provide a final and binding decision on a particular issue. Expert determinations are usually private contracts not susceptible to judicial review of matters such as not following the Rules of Natural Justice, but the expert does have a duty to observe procedural fairness as dictated by any agreed terms of reference, procedures or adopted code of conduct. Agreed terms of reference might include that the expert can have separate private conversations with individual Parties; be inquisitorial; investigate and perhaps not be obliged to give the results of those investigations to the Parties before making a decision; provide a decision binding on the Parties; award interest; proceed even if one of the Parties at some stage refuses to cooperate; and be indemnified by the Parties. Terms of reference may also include that liability for the payment of fees will be joint and several, that there can be a lien on the decision and so on.

There can thus be marked similarities between an expert determination and an adjudication, particularly where in the latter case the Parties agree both to the Adjudicator and to the adjudication decision being final and conclusive. Indeed the courts have suggested as much (*Bouygues* v *Dahl-Jensen* [1999 and 2000]).

It is therefore considered that the essential differences between expert determination and adjudication should have been clarified so that the drafters' causes for concern could be understood. On the face of it, this part of Paragraph 1.4 is not considered to be particularly helpful, and some might say it is unnecessary.

Paragraph 1.5 Binding decision

> *The Adjudicator's decision shall be binding until the dispute is finally determined by legal proceedings, by arbitration (if the contract provides for arbitration or the Parties otherwise agree to arbitration) or by agreement.*

A contractual adjudication clause such as this has still to be interpreted by the courts in relation to adjudication. By analogy with the judgments in the statutory Scheme adjudication cases, it is likely that by this paragraph the Parties are agreeing that, even if the subject of a major error of fact or a minor procedural defect, any decision of the Adjudicator must be regarded as temporarily binding and observed (usually by payment of any award) pending final determination. Of course, in contractual adjudications, just as in statutory adjudications, a jurisdictional error, bad faith or the like is likely to remove the binding nature of any decision. Furthermore, at the same time as being 'temporarily' binding, the Parties can between them (either before or after any decision) agree that the decision of the Adjudicator shall be 'finally' binding, or it can become so by default (e.g. see Clause 66(9)(b)).

One of the ways both Parties could 'by agreement' alter an adjudication decision is by having the same or another Adjudicator 'open up, review and revise' it in accordance with Paragraph 5.3 (see later) and agreeing to accept that second opinion. (It will be recalled that involvement with a previous adjudication decision is expressly excluded by the Scheme under Paragraph 9 (2), where it is an obligatory resignation issue!) Associated issues are discussed at length later within Paragraph 5.3.

Alternatively the Parties could both 'by agreement' abide by the result of another ADR method or to just agree a different decision between them.

Whether there is arbitration or litigation would depend upon the exact terms of the underlying contract. It should be noted that subsequent arbitration or litigation under this Paragraph would not be an appeal against the decision, but a consideration of the dispute as if the earlier adjudication decision had never been made. It will also be noted that there appears to be no bar to the decision of the Adjudicator being cited in evidence.

Paragraph 1.6 Implementation of decision without delay. Payment

The Parties shall implement the Adjudicator's decision without delay whether or not the dispute is to be referred to legal proceedings or arbitration.

Payment shall be made in accordance with the payment provisions in the Contract, in the next stage payment which becomes due after the date of issue of the decision, unless otherwise directed by the Adjudicator or unless the decision is in relation to an effective notice under Section 111 (4) of the [Construction] *Act.*

A Party cannot delay complying with a decision until it is decided later by arbitration or by the courts. The courts (e.g. *Macob* v *Morrison* [1999] and *Outwing* v *Randell* [1999]) support this basic principle where possible. To enforce compliance, the other Party can make specific use of Paragraph 6.7 of the ICE Procedure (see later) and seek summary enforcement in the courts.

Paragraph 2.1 Notice of Adjudication and contents

Any Party may give notice at any time of its intention to refer a dispute arising under the Contract to adjudication by giving a written Notice of Adjudication to the other Party. The Notice of Adjudication shall include:

- *(a) the details and date of the Contract between the Parties;*
- *(b) the issues which the Adjudicator is being asked to decide;*
- *(c) details of the nature and extent of the redress sought.*

A development in statutory Scheme adjudication has been the importance which the courts have come to attach to clarifying what the dispute is and the extent to which it is correctly reflected in the Notice of Adjudication. How such judgments will apply to contractual adjudications is not yet clear. There appears to be greater scope in a contractual adjudication for the Parties to agree between themselves and the Adjudicator any adjustment or enhancement of the description of the dispute being referred for adjudication. Notwithstanding any such agreement it is suggested however, that, even those involved in contractual adjudications should attempt to make the 'dispute' stated in the Notice of Adjudication clear.

Paragraph 3.1 Pre-named, pre-agreed Adjudicator

> *Where an Adjudicator has either been named in the Contract or agreed by the Parties prior to the issue of the Notice of Adjudication, the Party issuing the Notice of Adjudication shall at the same time send to the Adjudicator a copy of the Notice of Adjudication and a request for confirmation, within four days of the date of issue of the Notice of Adjudication, that the Adjudicator is able and willing to act.*

i.e. any pre-named or otherwise pre-agreed Adjudicator shall be sent the Notice of Adjudication at the same time as the other Party and within 4 days must confirm its willingness to act.

If an Adjudicator does not respond in the affirmative until, say, day 5 or 6, on the face of this Paragraph it is too late, even if that Adjudicator is the one preferred by both Parties. The Referring Party is obliged to go on to the next appointment stage, notwithstanding that this results in an overall delay in the selection of an Adjudicator, and one neither Party would prefer to have. This appears unnecessarily onerous and could surely have been left more open by using some minor rewording, such as adding 'unless the referring Party agrees otherwise.' In practical terms, if both Parties agree, another option to proceeding to Paragraph 3.3 could be to reissue the Notice of Adjudication and give the potential Adjudicator another four days.

Paragraph 3.2 Referring Party to provide proposed Adjudicator(s)

> *Where an Adjudicator has not been so named or agreed, the Party issuing the Notice of Adjudication may include with the Notice the names of one or more persons with their addresses who have agreed to act, any one of whom would be acceptable to the referring Party, for selection by the other Party.* [Where there is more than one person] *The other Party shall select* [one of them] *and notify the referring Party and the selected Adjudicator within four days of the date of issue of the Notice of Adjudication.*

There could be a slight clarification of the wording [as indicated in square brackets] to cover more than one person. A procedure where it appears that the Referring Party can, in effect, unilaterally impose an Adjudicator of its choice on the other Party (or produce a list of its own nominee Adjudicators from which the other Party has to make a selection), seems unlikely in many circumstances to meet with approval by that latter Party. How many times will a referring Contractor, sub-Contractor or Employer respectively do anything but suggest another Adjudicator experienced as a Contractor, sub-Contractor or Employer to determine the dispute?

Paragraph 3.3 Appointment by named body or ICE

> *If confirmation is not received under paragraph 3.1 or a selection is not made under paragraph 3.2 or the Adjudicator does not accept or is unable to act then either Party may within a further three days request the person or body named in the Contract or if none is so named The Institution of Civil Engineers to appoint the Adjudicator. Such request shall be in writing on the appropriate form of application for the appointment of an Adjudicator and accompanied by a copy of the Notice of Adjudication and the appropriate fee.*

Thus, despite the apparent obligatory nature of the Paragraph 3.2 selection process, the Responding Party is not after all forced to select from a list (which might consist of one person) put together unilaterally by the Referring Party.

The wording of the foregoing would benefit more than most other Paragraphs from a few simple insertions and commas (underlining indicates words that could be omitted), e.g.:

> *If confirmation is not received under paragraph 3.1, or a selection is not made under paragraph 3.2, or the* [potential] *Adjudicator* [under paragraph 3.1 or 3.2] *does not accept or is unable to act, then either Party may within a further three days request the* [appointing] *person or* [appointing] *body named in the Contract, or if none is so named The Institution of Civil Engineers, to appoint the Adjudicator.* <u>*Such request*</u> [The latter] *shall be in writing on the* <u>*appropriate*</u> *form* <u>*of*</u> ['A]*pplication for the* [selection/] *appointment* <u>*of an*</u> [A]*djudicator' and accompanied by a copy of the Notice of Adjudication and the appropriate fee.*

It is not known where in the ICE Conditions the named person or body should be stated, if it is not to be an ICE appointee.

Paragraph 3.4 Obligation to use Adjudicator's Agreement and Schedule. Adjudicator's entitlement to reasonable fees and expenses. Need for signature

> *The Adjudicator shall be appointed on the terms and conditions set out in the attached Adjudicator's Agreement and Schedule and shall be entitled to be paid a reasonable fee together with his expenses. The Parties shall sign the agreement within 7 days of being requested to do so.*

This single Paragraph deals with three separate issues and would benefit from being expressed that way, i.e.

> *The Adjudicator shall be appointed on the terms and conditions set out in the attached Adjudicator's Agreement and Schedule[.]*

Why there should be an obligation for an Adjudicator to accept the ICE Procedure drafter's view of what are acceptable terms and conditions is also discussed in more detail in Chapter 9 dealing with AdjudicatorAappointments.

> [The Adjudicator] *shall be entitled to be paid a reasonable fee together with his expenses.*

This is self-evident if the adjudication process is to work.

> *The Parties shall sign the* [A]*greement within 7 days of being requested to do so.*

The latter is straightforward in its intent, i.e. the requirement for signatures provides reassurance to an Adjudicator that it has a valid contract relationship with the Parties. However, from a practical point of view, there is no fall-back provision for when one Party (or both Parties) does not, or will not sign, and what an Adjudicator should do in the circumstances. These matters are also discussed in Chapter 9.

It would have been better if the Paragraph expressly enabled the Adjudicator to, say, record any non-signing to both Parties and then to proceed regardless. The situation is then no different from statutory Scheme adjudications that do not have pro forma Agreements and thus have no need for signing them. There would still remain the possibility of an argument over whether the appointment was valid, and, importantly for the Adjudicator, there might be uncertainty regarding the Adjudicator's indemnity. However, the latter is contained within the Adjudication Procedure that is part of the ICE Conditions , not the Agreement. Therefore, as long as it can be established that an appointment is valid (and the Adjudicator is not acting outside its jurisdiction), despite no signature(s), it would appear that the indemnity (however limited – see later Paragraph 7.2) would also be in place.

Such potential problems would not arise for pre-selected named Adjudicators if it is made a requirement of the Parties, as a prerequisite for that Contract to become valid, to sign at the outset any Agreement(s) incorporated into the ICE Conditions for those pre-selected Adjudicator(s).

Paragraph 3.5 Selection of replacement Adjudicator

> *If for any reason whatsoever the Adjudicator is unable to act, either Party may require the appointment of a replacement Adjudicator in accordance with the procedure in paragraph 3.3.*

As Paragraph 3.3 incorporates by reference Paragraphs 3.1 and 3.2, it might be argued that the Parties are, for instance, free to agree a replacement Adjudicator between them, and why not? However, another interpretation might be that the Parties are obliged to use a third party organisation. This Paragraph would benefit from clarification.

Paragraph 4.1 Statement of case

> *The referring Party shall within two days of receipt of confirmation under* [paragraph] *3.1, or notification of selection under* [paragraph] *3.2, or appointment under* [paragraph] *3.3 send to the Adjudicator, with a copy to the other Party, a full statement of his case which should include:*
>
> *(a) a copy of the Notice of Adjudication,*
>
> *(b) a copy of any adjudication provision in the Contract, and*
>
> *(c) the information upon which he relies, including supporting documents.*

It will be noted that the statement of case includes the Notice of Adjudication, the minimum contents of which are itemised in earlier Paragraph 2.1.

There are a number of immediately foreseeable practical problems. For example, what does an Adjudicator do if the Referring Party takes longer than 2 days, or only sends some information, or an Adjudicator receives the documents promptly but, intentionally or otherwise, there is a delay in the other Party receiving its copies?

It could be argued that there are too many potential variations to be covered by what are intended as simple rules for a simple dispute resolution procedure. However, where obvious problems are not addressed, individual Adjudicators will have to try to deal with them later, with a probable lack of consistency of approach. It is considered that there should have been some attempt to deal with the more obvious situations that are likely to occur with some frequency.

The Construction Act merely says that the adjudication must have a timetable with the object of achieving an appointment in 7 days. The 2-day requirement stated in Paragraph 4.1 is a detail required by the ICE Procedure, not the Construction Act. The ICE Procedure could just as readily been worded more loosely whilst still complying with the 7-day target timetable, e.g. it could have added 'within two days, or, failing that, as soon as reasonably possible in the opinion of the Adjudicator'.

It might alternatively have added 'In the event of a short delay in the opinion of the Adjudicator, albeit one in excess of the two days allowed, equally applying to the Adjudicator and responding Party, the Adjudicator may proceed as if the documents had been received in two days'.

It is interesting that there is no reference to the Engineer's decision (or to any preceding Conciliator's recommendation) being involved in the statement of case. It is not excluded but surely warrants express inclusion? It could be argued that the adjudication is not on the Engineer's decision (or a Conciliator's recommendation) but on the underlying dispute, so there is no need to have that information. But how else can an Adjudicator confirm to its satisfaction that, for instance, a matter of dissatisfaction has gone through all the relevant ICE Clause 66 processes (or that the Conciliator's recommendation had not become final as a consequence of a too late serving of a notice of adjudication)?

These types of issues are discussed in more detail in Chapter 9.

Paragraph 4.2 Date of referral

The date of referral of the dispute to adjudication shall be the date upon which the Adjudicator receives the documents referred to in paragraph 4. 1. The Adjudicator shall notify the Parties forthwith of that date.

No matter how late an Adjudicator receives the Referring Party's full statement of case documents, the 28-day, or extended, timetable does not commence until it does. Some commentators might argue that it is a mandatory requirement of the Construction Act to refer a dispute within 7 days. If that is so, this Paragraph of the ICE Procedure could be none Construction Act compliant. Other commentators say the 7 day period should merely by an achievable objective which may sometimes not be met. This is discussed further under the heading of 'Urgency of appointment' in Chapter 9.

Paragraph 5.1 Decision in 28 days, or longer as agreed

The Adjudicator shall reach his decision within 28 days of referral, or such longer period as is agreed by the Parties after the dispute has been referred. The period of 28 days may be extended by up to 14 days with the consent of the referring Party ... .

The 28 days complies with the Construction Act and in isolation the wording is perfectly clear.

The use of the word 'reach' is noted. It will be seen later in comments against Paragraph 6.9 that greater clarity regarding the meaning of the words 'notify' and 'notification' would be helpful in Paragraphs 6.1, 6.3, 6.4 and 6.9. There is an associated possible (if not probable for many Adjudicators) requirement for payment of an Adjudicator's fees and expenses before releasing the decision. It is therefore suggested that it would have been very appropriate after the word 'reach' in this Paragraph to insert the words 'but not necessarily release, if notice has been given under Paragraph 6.6'.

... or such longer period as is agreed by the Parties.

This also conforms to the Construction Act. It will be noted that the Adjudicator must agree to any extension.

Paragraph 5.2 Matters to be dealt with by Adjudicator

The Adjudicator shall determine the matters set out in the Notice of Adjudication, together with any other matters which the Parties and the Adjudicator agree should be within the scope of the adjudication.

In law, an Adjudicator must reach a proper decision on whatever is validly placed before it. This clearly includes the matters set out in the Notice of Adjudication. The dispute initially referred to the Adjudicator can then be widened, but only if both the Parties and Adjudicator agree, i.e. additional matters cannot be introduced unilaterally by one Party unless the other Party agrees. This might include issues about which there is no current dispute (e.g. those that are still at the initial claim or claim negotiation stage, those that the other Party might not have seen or had an earlier opportunity to respond to or, in the ICE Conditions, those that have not been submitted for an Engineer's decision). In practice, an Adjudicator might find that there are several practical reasons for not agreeing with such a proposal (e.g. the timetable is too limited, there may be no spare diary space available to the Adjudicator even if the timetable is extendable by the Parties, the dispute would be widening outside the Adjudicator's area of expertise, the adjudication process is one that the Adjudicator wants to terminate as soon as possible due to the general uncooperative nature of the Parties and so on).

Paragraph 5.3 Adjudicator may open up any certificate etc. A decision cannot alter the right to vary the Conditions of Contract or the Works

> *The Adjudicator may open up review and revise any decision (other than that of an Adjudicator unless agreed by the Parties), opinion, instruction, direction, certificate or valuation made under or in connection with the Contract and which is relevant to the dispute.*
>
> *He may order the payment of a sum of money, or other redress but no decision of the Adjudicator shall affect the freedom of the Parties to vary the terms of the Contract or the Engineer or other authorised person to vary the Works in accordance with the Contract.*

The first sentence of this Paragraph allows an Adjudicator to open up disputes that have been the subject of previous contractual adjudication decisions, including its own. Section 108 (3) of the Construction Act requires an adjudication decision to be binding until, *inter alia*, 'the dispute is finally determined ... by agreement'. This Paragraph expands on the requirements of Section 108 (3) by expressly including that the Parties (i.e. this must be a joint decision, not one taken by just one Party) can agree to revise a previous adjudication decision.

However, an Adjudicator cannot overturn a previous adjudication decision expressly stated by the Parties as being a final determination of a dispute. It will be noted that Section 108 (3) of the Construction Act expressly allows a decision to be a final determination if both Parties agree. Also, Clause 66 (9)(b) of the Conditions of Contract states that an adjudication decision will become final and binding if no Notice to Refer to arbitration is served within three months of receipt of that decision.

Other decisions that cannot be overturned include those arising under other contractual procedures such as expert determinations, conciliations, mediations and so on, which have been agreed to be final and binding.

Paragraph 5.3 applies to matters arising 'under or in connection with' the contract (such as contractual conciliation requirements). If expert determination, conciliation or any other final and conclusive agreement had been concluded privately outside the contract, it is probable that the courts would decide that as they themselves cannot normally reopen such matters, then they should not be reopened up by an Adjudicator either. The latter is discussed in more detail in Chapter 14 on Miscellaneous Issues.

In summary therefore, until decided by the courts where necessary, it is suggested that it would be prudent for an Adjudicator to assume that:

(a) It *cannot* 'open up, review' etc. 'any certificate, decision' etc. agreed by both Parties (either within or outside an underlying contract) to be final and conclusive, or expressly stated (within an underlying contract) to become final and conclusive by default if certain defined circumstances do not occur, notwithstanding that the adjudication procedures may be stated to take precedence over underlying contract clauses.
(b) It *can* open any other 'certificate, decision' etc. under this Paragraph, including previous adjudication decisions as long as any other requirements are met, e.g. that to reopen an adjudication needs the agreement of both Parties.

Regarding the second part of Paragraph 5.3, this relates to opening up of decisions, certificates and so on 'under or in connection with' the Contract and which is relevant to the dispute.

This widens the scope of Section 108 (1) of the Construction Act beyond matters arising 'under' the construction contract to cover issues such as misrepresentation or tort. A contract clause can, of course, do this as long as the minimum requirements of the Construction Act are covered.

However, Paragraphs 1.2 and 2.1 of the ICE Procedure refer only to disputes arising 'under the Contract' and the ICE Conditions Clause 66 (6)(a) refers to 'under the contract'. Referring again to Paragraph 1.1 of the ICE Procedure, which says that the ICE Procedure takes precedence over the contract, the latter ICE requirement is overruled by the wider scope of Paragraph 5.3, but not the other ICE Procedure Paragraphs. This all seems unnecessarily confusing and contradictory.

The issue of 'under' or 'in connection with' and similar phrases is discussed in detail in Chapter 2, under Section 108 (1).

Paragraph 5.4 Time limit for response

The other Party may submit his response to the statement under paragraph 4.1 within 14 days of referral. The period of response may be extended by agreement between the Parties and the Adjudicator.

Recognising that a Responding Party will not always have a response, the words 'shall, if it has a response' may have been better here, instead of 'may'. If so, and a response was not forthcoming despite the positive obligation to respond within a stated time, an Adjudicator would be more justified in taking a decision to proceed *ex parte.*

Words such as 'or earlier if instructed by the Adjudicator' could also have been added after the end of the sentence, in order to give the Adjudicator an option to ask for a response earlier.

Paragraph 5.5 Adjudication at Adjudicator's complete discretion

The Adjudicator shall have complete discretion as to how to conduct the adjudication, and shall establish the procedure and timetable, subject to any limitation that there may be in the Contract or the Act. He shall not be required to observe any rule of evidence, procedure or otherwise, of any court. Without prejudice to the generality of these powers, he may:

- *(a) ask for further written information;*
- *(b) meet and question the Parties and their representatives;*
- *(c) visit the site;*
- *(d) request the production of documents or the attendance of people whom he considers could assist;*
- *(e) set times for (a) – (d) and similar activities;*
- *(f) proceed with the adjudication and reach a decision even if a Party fails:*
  - *(i) to provide information;*
  - *(ii) to attend a meeting;*
  - *(iii) to take any other action requested by the Adjudicator;*
- *(g) issue such further directions as he considers to be appropriate.*

Paragraph 5.6 Legal and technical advice

The Adjudicator may obtain legal or technical advice having first notified the Parties of his intention.

This seems straightforward – if the Parties are expected to pay for something they should be told in advance. The Adjudicator should pass on to both Parties the advice received.

Paragraph 5.7 Joining another Party

Any Party may at any time ask that additional Parties shall be joined in the Adjudication. Joinder of additional Parties shall be subject to the agreement of the Adjudicator and the existing and additional Parties. An additional Party shall have the same rights and obligations as the other [existing] *Parties, unless otherwise agreed by the Adjudicator and the* [existing and additional] *Parties.*

The joining of other Parties is dealt with in some detail in Chapter 14.

Paragraph 6.1 Notification of reaching a decision. Reasons

The Adjudicator shall reach his decision and so notify the Parties within the time limits in paragraph 5.1 and may reach a decision on different aspects of the dispute at different times.

The power to reach a decision on different aspects of the dispute at different times can be very useful and may affect how the rest of the adjudication proceeds.

He shall not be required to give reasons.

This reduces the likelihood of appeals on the Adjudicator's decision by not requiring the Adjudicator (unless it wants to) to provide reasons in its decision. Reasons are discussed in more detail in Chapter 13 dealing with the Decision.

Paragraph 6.2 Interest

The Adjudicator may in any decision direct the payment of such simple or compound interest at such rate and between such dates or events as he considers appropriate.

Under common law, there is no power to award interest, so this has to be expressly given, as it is here. It will be noted that there is no stated requirement that there must be a claim for interest, before an Adjudicator can award it (see also Interest in Chapter 13).

Paragraph 6.3 Late decision, 7 days notice of replacement of Adjudicator

Should the Adjudicator fail to reach his decision and notify the Parties in the due time [,] *either Party may give seven days notice of its intention to refer the dispute to a replacement Adjudicator* [who shall be] *appointed in accordance with the procedures in paragraph 3.3.*

Paragraph 6.4 Decision remains effective unless dispute has previously been referred to a replacement Adjudicator. Adjudicator then forfeits entitlement to fees

If the Adjudicator fails to reach and notify his decision in due time but does so before the dispute has been referred to a replacement Adjudicator under paragraph 6.3 his decision shall still be effective.

If the Parties are not so notified then the decision shall be of no effect and the Adjudicator shall not be entitled to any fees or expenses but the Parties shall be responsible for the fees and expenses of any legal or technical adviser appointed under paragraph 5.6 subject to the Parties having received such advice

In other words, taking Paragraph 6.3 and 6.4 together, without reaching any agreement with the Parties for an extension, an Adjudicator has at least 7 days more than the 28-day, or extended, timetable requirement to reach and notify its decision and not forfeit an entitlement to be paid its fees and expenses. This right of an Adjudicator to unilaterally have a 7-day extension would not appear to be in accordance with the tight timetable intention of the Construction Act, even if it is a more preferable and practical option than having to start all over again. It is certainly a useful practical fall-back position if an Adjudicator is, say, ill or otherwise unavoidably delayed for a short time, although it is possible, of course, that some Adjudicators will make regular use of this when organising their diaries. But interestingly, why limit the Adjudicator to 7 days? Why not 14 days or a month? Is it the maximum the drafters thought they could get away with without lawyers for one Party or another using it successfully as an excuse for appealing the decision due to it not complying with at least the spirit of the Construction Act?

However, this is without doubt a valuable provision. Many standard conditions of contract are silent as to subsequent procedure if a decision does not appear on the 28th day, leaving it to the Parties to commence another adjudication. This procedure at least allows a decision to be received later and, if not, at least provides some continuity. It could perhaps be further improved by compelling the Parties to send to the replacement Adjudicator copies of all submissions already made by the Parties during the previous adjudication so as to speed up progress towards a decision.

Paragraph 6.5 Parties' costs, Adjudicator's fees and expenses

The Parties shall bear their own costs and expenses incurred in the adjudication.

Parties shall be jointly and severally responsible for the Adjudicator's fees and expenses, including those of any legal or technical adviser appointed under paragraph 5.6, but in his decision the Adjudicator may direct a Party to pay all or part of his fees and expenses. If he makes no such direction the Parties shall pay them in equal shares.

Under this Paragraph there is an express requirement for the Parties to pay their own costs.

The Adjudicator may decide that one Party should pay more than half of its fees and expenses, because, for instance, that Party has been obstructive and uncooperative during the adjudication process. The Adjudicator must be sure that it is allowed to do this, i.e. that it lies within its jurisdiction to do so, otherwise a Party may consider that it has grounds for an appeal, on that basis or alternatively on the basis of partiality. However, where an Adjudicator has not given reasons (as permitted in Paragraph 6.1) the Parties may find it difficult to launch a successful appeal.

Paragraph 6.6 Release of decision after payment of fees

At any time until 7 days before the Adjudicator is due to reach his decision, he may give notice to the Parties that he will deliver it only on full payment of his fees and expenses. Any Party may then pay these costs in order to obtain the decision and recover the other Party's share of the costs in accordance with paragraph 6.5 as a debt due.

It is suggested that this is best stated by an Adjudicator immediately on appointment, particularly when, for some Parties, it is possible that time might be needed to raise the necessary amount of money.

As will be seen later in the discussions on the ICE Schedule to the Adjudicator's Agreement, the Adjudicator is also entitled to ask for an Advance Payment. This would provide security for all, or part, of its fees and expenses. The Adjudicator might therefore not wish to take advantage of Paragraph 6.6 in such circumstances.

It is noted that some commentators are not in favour of provisions such as those given in this Paragraph. It can be assumed that they are generally not Adjudicators themselves!

Paragraph 6.7 Summary enforcement. Decision not to be referred to another Adjudicator

The Parties shall be entitled to the relief and remedies set out in the decision and to seek summary enforcement thereof, regardless of whether the dispute is to be referred to legal proceedings or arbitration.

As described earlier under Clause 66 (9)(a), enforcement (summary or otherwise) will be by application to the courts. In fact, the courts have stated (*Macob* v *Morrison* [1999]) that the usual procedure in England and Wales following an adjudication decision, particularly in money cases, would be for a Party to issue court proceedings claiming the amount due, followed by an application to the courts for summary judgment, all in accordance with normal debt recovery procedures. This procedure could also be enhanced by a Party requesting a court to reduce the normal timetable for debt recovery, e.g. the intimation period (the period in which a defendant must reply if it wishes to defend an action) (*Outwing* v *Randell* [1999]). Summary enforcement is discussed in more detail in Chapter 13 dealing with the Decision.

No issue decided by an Adjudicator may subsequently be laid before another Adjudicator unless so agreed by the Parties.

It is not clear why this sentence is added to this Paragraph, being an entirely different issue to that stated in the first part of it and furthermore a subject already included within Paragraph 5.3, as discussed earlier.

Paragraph 6.8 Parties subsequent rights and obligations not affected

In the event that the dispute is referred to legal proceedings or arbitration, the Adjudicator's decision shall not inhibit the court or Arbitrator from determining the Parties' rights or obligations anew.

The purpose of this Paragraph is presumably to make clear that an Arbitrator or court will not in these circumstances be assessing the correctness or otherwise of an adjudication decision, but will instead consider the dispute anew.

It is noted that although the Adjudicator cannot be a witness in such proceedings (see Paragraph 7.1) there is no stated bar to the disclosure of the Parties' previous submissions to the Adjudicator or of the Adjudicator's decision. Either Party can presumably present these to the tribunal and this might lead to subsequent proceedings being significantly shortened.

Paragraph 6.9 Correction of clerical mistakes

> *The Adjudicator may on his own initiative, or at the request of either Party, correct a decision so as to remove any clerical mistake, error or ambiguity provided that the initiative is taken, or the request is made within 14 days of the notification of the decision to the Parties. The Adjudicator shall make his corrections within 7 days of any request by a Party.*

i.e. the Adjudicator can, if it wants to, correct any clerical mistake, error or ambiguity in a decision within 14 days of notification if the Adjudicator spots it, or must make the correction within 7 days if one of the Parties asks for it to be done.

Aspects of the correction of errors in adjudication decisions, including some relevant court decisions, are discussed under the heading 'Correction of Errors' within Chapter 13.

The Paragraph requires an Adjudicator to make amendments within 7 days of matters requiring amendment being pointed out by a Party, the Party having 14 days from *notification* in which to do so, as already described.

However, whereas Paragraph 6.1 requires the Adjudicator to reach a decision and *notify* the Parties (presumably of the decision's completion but this is not stated) within 28 days (or longer if there is an agreed extension), under Paragraph 6.6 the Adjudicator will probably require its fees to be paid before *release* of the decision. If there is any delay in the Adjudicator receiving its fees, and hence releasing its decision, that will seriously impinge on any Party's time (which is measured from the *notification* date, not from the date of release) to note any adjustment they might wish the Adjudicator to make, at the extreme there being no time available at all! This cannot be considered reasonable and Paragraph 6.9 needs amendment, e.g. by altering 'notification' to 'receipt'.

It is also suggested that the wording of Paragraphs 6.1, 6.4, 6.6 and 6.9 would also be clearer if the words 'notify' and 'notification' were amplified to 'notification of availability, subject to any order for the prior receipt of the Adjudicator's fees and expenses under Paragraph 6.6', or similar.

Paragraph 7.1 Adjudicator not to be subsequent witness or Arbitrator

> *Unless the Parties agree, the Adjudicator shall not be appointed Arbitrator in any subsequent arbitration between the Parties under the Contract.*

No Party may call the Adjudicator as a witness in any legal proceedings or arbitration concerning the subject matter of the adjudication.

It is only sensible that the Adjudicator should not be the subsequent Arbitrator (as, human nature being what it is, this is likely to effectively tend towards the same decision being reached twice). In case this arises accidentally, say by an overlooked contractual provision, this Paragraph precludes that happening. However, the nexus appears to apply not just to the dispute in hand but to the entire contract, although that, too, may be sensible if the Adjudicator has, for example, declared a view as to the conduct of one or other of the Parties.

Adjudicators cannot be called as witnesses. This preserves the confidentiality of their deliberations and avoids them spending significant time (perhaps days) in court waiting rooms, providing statements and evidence about their decision, probably at court fee rates. This bar on an Adjudicator being called as a witness does not, however, prevent adjudication decisions or the submissions made by the Parties during adjudications being introduced in court.

Paragraph 7.2 No liability except for bad faith. Indemnity

The Adjudicator shall not be liable for anything done or omitted in the discharge or purported discharge of his functions as Adjudicator unless the act or omission is in bad faith, and any employee or agent of the Adjudicator shall be similarly protected from liability.

The Parties shall save harmless and indemnify the Adjudicator and any employee or agent of the Adjudicator against all claims by third parties and in respect of this shall be jointly and severally liable.

The latter means that an Adjudicator can therefore pass all third party claims against it directly to the Parties to sort out. Without this indemnity, even if ultimately found not to be liable, an Adjudicator might have to spend considerable non-productive time significantly eroding limited profit margins in order to assist extricate itself from proceedings. In addition, this might unreasonably delay enforcement and provide a means of reviewing and perhaps reopening the decision.

Even now, neither of the above provides total Adjudicator immunity (even where an Adjudicator has not acted in bad faith).

This is firstly because there is no specific reference to negligence, the most likely claim against an Adjudicator, and there therefore may be no immunity against a claim for it. It is considered that an additional express reference to negligence should have been provided for absolute certainty of cover. The only other way to cover this objective would have been in the Adjudicator's Agreement, but as discussed in Paragraphs 1.1 and 3.4, amendments to this are not theoretically possible with pro-forma compulsory wording. The issue is discussed under Section 108 (4) in Chapter 2 on the Construction Act.

Secondly, the Parties (and the Adjudicator if not adequately indemnified against all possible actions by third parties in the Adjudication Procedure or in the Adjudicator's Agreement) might have potential liabilities under the Construction (Rights of Third Parties) Act 1999. This needs wording such as 'for the avoidance of doubt in relation to the Construction (Rights of Third Parties) Act 1999, the Parties to this Adjudication do not intend that any of its terms or those in the Adjudicator's Agreement shall be enforceable against the Adjudicator (and the Parties) by any third party.'

Paragraph 7.3 ICE disclaimer

Neither The Institution of Civil Engineers nor its servants or agents shall be liable to any Party for any act omission or misconduct in connection with any appointment made or any adjudication conducted under this Procedure.

Paragraph 7.4 Recorded delivery

All notices shall be sent by recorded delivery to the address stated in the Contract for service of notices, or if none, the principal place of business or registered office (in the case of a company).

It is assumed that this means a formal service operated by a postal/courier organisation and excludes any other means that results in a record of receipt being obtained, e.g. including delivery by hand directly between the Parties.

In practice, whilst key documents such as the statement of case and any responses are normally sent by some form of recorded delivery, most other communications are by e-mail or fax perhaps, backed up by an original sent by first class post. It might be better if this Paragraph recognised such everyday realities. (It will be noted that the JCT Procedure, Paragraph 4.2, deals with this issue better – see next chapter.)

The address provisions also appear unnecessarily prescriptive, e.g. the addresses for those dealing with the dispute might be somewhere other than those stated in the contract. This could have been made more realistically flexible by using, for example, the insertion of the words 'unless advised otherwise' after '(in the case of a company)'.

It is suggested that the Adjudicator's Agreement should be the place to state categorically the address to which the Adjudicator wants documents sent, and perhaps a telephone and fax number. For example, an employee, director or partner of an organisation acting as named Adjudicator (Paragraph 1.3) might operate from home for the purposes of the adjudication, but the initial Notice of Adjudication, acceptance of the appointment and place of receiving the Referral Documents might have made use of the organisation address.

Any agreement required by this Procedure shall be evidenced in writing.

It is not understood what is meant by this and why it is here within a Paragraph dealing with the delivery of notices, unless there is a typographical error or a separate Paragraph number missing. For example, it could be dealing with the Adjudicator's Agreement. If correct, 'agreement' should read 'Agreement' and the latter should be defined in Paragraph 8. Another, more probable, view is that it is a general provision concerned with any agreements reached during the adjudication process. If so, it would still benefit from better explanation and a separate Paragraph number.

Paragraph 7.5 Law

This Procedure shall be interpreted in accordance with the law of the Contract.

Paragraph 8 Definitions

It is suggested that the following definitions could usefully be added:

- 'Adjudicator's Agreement'
- 'Appointment Fee'. (See later in the short discussion on the Schedule to the Adjudicator's Agreement.)

Paragraph 9.1 ICE Small Works Consultancy Agreement

When this Procedure is used with The Institution of Civil Engineers' Agreement for Consultancy Work in Respect of Domestic or Small Works, the Adjudicator may determine any dispute in connection with or arising out of the Contract.

It will be noted that this Paragraph applies to disputes 'in connection with' as well as 'arising out of' the dispute. The issues of 'in connection with' and 'arising out of' have been dealt with earlier in this chapter.

Schedule to the Adjudicator's Agreement

Advance payment

Matters relating to Adjudicator appointments generally, including possible terms and conditions, are discussed in Chapter 9. They equally apply to the limited issues listed in the ICE Adjudicator's Schedule. However, of particular note and not mentioned in either Clause 66 of the ICE Conditions or in the ICE Procedure, is that an Adjudicator can request an 'appointment fee', in other words, an advance payment. As previously noted, this could usefully have been defined in Paragraph 8 Definitions.

It is not uncommon for an advance payment, possibly one roughly equivalent to the anticipated total fees expected, to be requested by Adjudicators. In view of the limited involvement and potential for profit involved in each adjudication dispute, the last thing an Adjudicator wants is to be involved in non-chargeable time chasing payment of bona fide fees and suffering delays in payment in the meantime. The amount of fees involved are likely to be so much less than the amounts in dispute, that the requirement for an advance payment should not be an overly onerous requirement on Parties in most circumstances. The Adjudicator will, at an early stage, be placed on warning of final fee payment problems ahead if it is.

The advance payment is payable equally by both Parties within 14 days of the Adjudicator being appointed. As is normal within the ICE Procedure, there is no stated fall-back position if either or both parts of the advance are not paid, and thus immediately there is a resulting potential problem for the Adjudicator, i.e. to proceed or to stop. Ways in which this problem could easily and reasonably be tackled are discussed in Chapter 9.

Chapter 4

JCT Adjudication Procedure

General

This chapter considers the Clause 41A Adjudication provisions [hereafter the JCT Procedure] contained within the JCT Local Authorities with Quantities Standard Form of Building Contract, 1998 Edition, incorporating Amendments 1:1999, 2:2000 and 3:2001. The basic provisions are the same or similar to other JCT Forms of Contract.

The JCT Procedure is based on what was originally issued as part of JCT80 Amendment 18, which came with Guidance Notes. It is assumed that the latter remain valid for those Paragraphs (and Articles) that are essentially unchanged between the Amendment 18 and the current JCT Procedures.

The following considers the detailed requirements. The JCT Procedure Paragraphs are quoted in full and given in italics.

Articles

The following Article paragraphs are generally paraphrased and reference to the original text will be needed to see the full wording in its correct context.

Article 5

> *Either Party may refer any dispute or difference arising under the contract to Clause 41A adjudication.*

The JCT Guidance Notes bound into the Amendment 18 document state that either Party must refer a dispute or difference by giving a notice [of adjudication] to the other Party.

Article 5 expressly refers to disputes arising 'under' the contract. The JCT Guidance Notes bound into the Amendment 18 document state that a Leading Counsel has advised that an Adjudicator could not therefore decide, for example, on the existence of a contract or a right to sue for misrepresentation, as these would not be arising 'under' the contract. It will be noted that this is narrower than the scope of 'under' and 'in connection with' stated in following Articles 7A and 7B.

This issue is considered in more detail in Section 108 (1) of Chapter 2 dealing with the Construction Act.

Article 7A

Subject to the Parties still retaining a right to Clause 41A adjudication, if arbitration under Clause 41B applies, other than enforcement of an adjudication decision, generally all disputes or differences both arising 'under' and 'in connection with' the contract must be resolved by arbitration.

If a Referring Party does not want to use adjudication to resolve a dispute or difference, the matter must be resolved by arbitration if there are arbitration provisions in the contract.

Article 7B

Subject to the Parties still retaining a right to Clause 41A adjudication, if arbitration under Clause 41B is excluded, generally all disputes or differences both arising 'under' and 'in connection with' the contract, must be resolved by legal proceedings.

If a Referring Party does not want to use adjudication to resolve a dispute or difference, and there are no arbitration provisions in the contract, the matter must be resolved by the courts.

Enforcement of an Adjudicator's decision must be by the courts, even when there are arbitration clauses in a contract. Matters relating to enforcement of Adjudicator's decisions are considered in more detail in Chapter 15.

The importance of getting these Articles correctly incorporated into the Parties' contract has already been illustrated by one adjudication case (*Finney* v *Vickers* [2001]). In that case, one Party's attempt, for whatever reason, to have an issue resolved by arbitration was prevented because the latter was expressly excluded under Clause 41B.

The JCT Procedure

For ease of reference in the following, Paragraph 41A.1, Paragraph 41A.2 etc. are referred to as Paragraph 1, Paragraph 2 etc. Also, in the commentary, 'dispute or difference' is abbreviated to 'dispute', and 'notice', 'notice of intention to refer to adjudication' or similar is changed to 'notice [of adjudication]'.

Paragraph 1 Application

Clause 41A applies where, pursuant to Article 5, either Party refers any dispute or difference arising under this Contract to adjudication.

What constitutes a dispute is dealt with in detail in Chapter 14.

Paragraph 2 Appointment of Adjudicator and referral within 7 days

The Adjudicator to decide the dispute or difference shall be either an individual agreed by the Parties or, on the application of either Party, an individual to be nominated as the Adjudicator by the person named in the Appendix ('the nominator'). Provided that

> *1 no Adjudicator shall be agreed or nominated under* [Paragraph] *2 or* [Paragraph] *3 who will not execute the Standard Agreement for the appointment of an Adjudicator issued by the JCT (the 'JCT Adjudication Agreement') with the Parties, and*
>
> *2 where either Party has given notice of his intention to refer a dispute or difference to adjudication then*
>
> > *– any agreement by the Parties on the appointment of an Adjudicator must be reached with the object of securing the appointment of, and the referral of the dispute or difference to, the Adjudicator within 7 days of the date of the notice of intention to refer (see* [Paragraph] *4.1);*
> >
> > *– any application to the nominator must be made with the object of securing the appointment of, and the referral of the dispute or difference to, the Adjudicator within 7 days of the date of the notice of intention to refer;*
>
> *3 Upon agreement by the Parties on the appointment of the Adjudicator or upon receipt by the Parties from the nominator of the name of the nominated Adjudicator the Parties shall thereupon execute with the Adjudicator the JCT Adjudication Agreement.*

Footnotes:

The nominators named in the Appendix have agreed with the JCT that they will comply with the requirements of clause 41A on the nomination of an Adjudicator including the requirement in [Paragraph] *2.2 for the nomination to be made with the object of securing the appointment of, and the referral of the dispute or difference to, the Adjudicator within 7 days of the date of the notice of intention to refer: and will only nominate Adjudicators who will enter into the 'JCT Adjudication Agreement'.*

Of particular relevance and usefulness is Paragraph 5.6 (see later) that expressly states that an adjudication decision will remain valid even if a Party does not 'execute the JCT Adjudicator Agreement'. This partly overcomes the potential difficulties of one Party not signing a pro forma Adjudicator Agreement, problems that are not addressed at all in other adjudication procedures.

Having a pro forma Adjudicator's Agreement does not prevent an Adjudicator from proposing additional terms and conditions, or even variations from what might be in the pro forma Agreement, adjudication procedures, or the Construction Act (e.g. the Parties can even agree to waive statutory requirements in individual circumstances as long as the waiver is clearly expressed). However, having made a proposal does not mean the same as having it accepted. At the commencement of an appointment, the Adjudicator can decide not to commence at all if its proposals are not met. If proposals are made during the adjudication and not met, the only options are for the Adjudicator to proceed anyway as if the proposal was not made, or to resign.

For example, there is nothing mentioned in the pro forma Adjudicator's Agreement about providing the Adjudicator with an Advance Payment as security for some, or all, of its costs. This does not mean that the Adjudicator cannot ask for an Advance Payment or that the Parties cannot agree to provide one. Equally, they are under no obligation to do so. The issue is one between the Parties and the Adjudicator and if the Adjudicator considers the matters important enough it can decide not to proceed with the appointment.

Another example is that there is nothing mentioned in the pro forma Adjudicator's Agreement about paying the Adjudicator its outstanding fees and expenses before it releases its decision. If raised at the time of the proposed appointment, the default option for the Adjudicator is not to proceed with the appointment. If proposed by the Adjudicator during the adjudication and not accepted by the Parties, the Adjudicator has no option but to resign or to continue, and, if it continues, to reach and release its decision within the stated Paragraph 5.3 timescale.

It is suggested that in making any proposals, the Adjudicator should make it clear, certainly to relatively 'unsophisticated' or legally unrepresented Parties, that they are under no obligation to agree to them. Otherwise it is quite possible that complaints might be made, not least to the Adjudicator's appointing body, if a Party feels it has been misled. Issues related to pro forma agreements are discussed in Chapter 9 dealing with Adjudicator Appointments.

The Guidance Notes, page 45, refers to the need to use a different pro forma Agreement for when the Parties name an Adjudicator in the Appendix to the contract. That pro forma must be signed by the Adjudicator and the Parties at the time of the contract. In this situation, it is stated in the Guidance Notes on Page 55 that the agreed person must be both an individual and also one who is not an employee or otherwise engaged by either Party, unless both Parties agree otherwise. Paragraph 2, which appears to be only concerned with the 'normal' pro forma Adjudication Agreement, does not therefore apply to the latter particular method of selecting an Adjudicator.

Paragraph 3 Replacement Adjudicator

> *If the Adjudicator dies or becomes ill or is unavailable for some other cause and is thus unable to adjudicate on a dispute or difference referred to him, then either the Parties may agree upon an individual to replace the Adjudicator or either Party may apply to the nominator for the nomination of an Adjudicator to adjudicate that dispute or difference; and the Parties shall execute the JCT Adjudication Agreement with the agreed or nominated Adjudicator.*

Even where there is a pre-named Adjudicator in the contract, the same replacement procedure can apply, i.e. a replacement can be agreed by the Parties, or either Party can apply to any nominator named in the Appendix, or if no nominator, to the President or Vice-President of the RIBA. It will be noted that the latter is an additional option to those stated in Paragraph 3 (unless the nominator and the RIBA President or Vice-President are one and the same).

Paragraph 4.1 Notice of adjudication, referral in 7 days, details to be supplied to other Parties

> *When pursuant to Article 5 a Party requires a dispute or difference to be referred to adjudication then that Party shall give notice* [the notice of Adjudication] *to the other Party of his intention to refer the dispute or difference, briefly identified in the notice, to adjudication.*
>
> *If an Adjudicator is agreed or appointed within 7 days of the notice then the Party giving the notice shall refer the dispute or difference to the Adjudicator ('the referral') within 7 days of the notice. If an Adjudicator is not agreed or appointed within 7 days of the notice the referral shall be made immediately on such agreement or appointment. The said party shall include with that referral particulars of the dispute or difference together with a summary of the contentions on which he relies, a statement of the relief or remedy which is sought and any material he wishes the Adjudicator to consider.*
>
> *The referral and its accompanying documentation shall be copied simultaneously to the other Party*

Once a Party has notified the other Party (presumably in writing) of its intention to refer a dispute to adjudication, there appears to be no particular urgency needed for the appointment of an Adjudicator. If the appointment is made within 7 days of the initial notification (assumed to be the date the other Party receives the written notification), the referral must be made to the Adjudicator within the same 7 day period. If the appointment is made any time later than that, the referral must be provided, to all intents and purposes, simultaneously with the appointment. This seems reasonable and clearly negates any need to recommence the adjudication process from the beginning if the 7 day timetable is not met, thus preventing further delay.

The JCT contract and the JCT Adjudication Agreement will often both be signed at the outset with few appointment difficulties. In practice, however, naming the Adjudicator in advance is falling out of favour in preference to the appointment of an Adjudicator most suited to the particular dispute at the time when the dispute arises. The JCT contracts allow for this. An area as yet untested in relation to adjudication is the extent to which the timing provisions of Section 108 (2)(b) of the Construction Act relating to the appointment of the Adjudicator is mandatory, directory or permissive. If the legislation is mandatory then an adjudication which has failed within the stated timescale to appoint an Adjudicator and refer the dispute may later be struck down as voidable. If, however, the legislation is merely directory, then a failure to do so will not invalidate an adjudication. The drafters of the JCT Procedure must hope that the legislation is directory because completing all of the procedures (including the Adjudicator's Agreement) and getting the dispute referred to the Adjudicator within 7 days will often be impractical. A day or so may not be fatal if the legislation is interpreted as directory, although a delay of a few days may be. The insistence of the JCT that their Adjudicator's Agreement be signed may, if tested in the courts, turn out to be a problem they have created for themselves (notwithstanding the terms of Paragraph 5.6 – see later). It is suggested that the most prudent course of action for Parties and Adjudicators in the meantime is to try to ensure the validity of their adjudications by doing their utmost to comply with the 7-days limit.

Paragraph 4.2 Methods of sending referral

> *The referral by a Party with its accompanying documentation to the Adjudicator and the copies thereof to be provided to the other Party shall be given by actual delivery or by FAX or by special delivery or recorded delivery. If given by FAX then, for record purposes, the referral and its accompanying documentation must forthwith be sent by first class post or given by actual delivery. If sent by special delivery or recorded delivery the referral and its accompanying documentation shall, subject to proof to the contrary, be deemed to have been received 48 hours after the date of posting subject to the exclusion of Sundays and any Public Holiday.*

The referral shall be by facsimile, actual, special or recorded delivery. If sent by facsimile, copies must also be supplied by first class post or actual delivery. If sent by special or recorded delivery, receipt will be assumed 48 hours later, subject to adjustment to take into account Sundays or public holidays.

Paragraph 5.1 Confirmation of receipt of referral by Adjudicator

> *The Adjudicator shall immediately upon receipt of the referral and its accompanying documentation confirm the date of that receipt to the Parties.*

This confirmation of the date received is in fact a notification that the 28-day (or later if agreed) timetable to reach a decision has commenced.

Paragraph 5.2 Response within 7 days

> *The Party not making the referral may, by the same means stated in* [Paragraph] *4.2, send to the Adjudicator within 7 days of the date of the referral, with a copy to the other Party, a written statement of the contentions on which he relies and any material he wishes the Adjudicator to consider.*

It is assumed that the 'date of referral' is the Paragraph 5.1 date of actual receipt by the Adjudicator, but it would have been clearer to have stated this.

Paragraph 5.3 28 days to reach and send decision unless longer agreed

> *The Adjudicator shall within 28 days of the referral under* [Paragraph] *4.1, and acting as an Adjudicator for the purposes of S.108 of the Housing Grants, Construction and Regeneration Act 1996 and not as an expert or an Arbitrator, reach his decision and forthwith send that decision in writing to the Parties.*

It is assumed that the timetable commences when the referral is received by the Adjudicator. To be an effective referral, the items included must, at that time, be those listed in Paragraph 4.1. It is further suggested that if they do not, then the date the last is received by the Adjudicator is the start date for the reference.

The decision must be reached and actually sent to the Parties within the stated time, e.g. there is no withholding it until payment of outstanding fees and expenses are paid. There is no obvious fall-back position if the decision is late by a few days. (In theory, the Parties could still agree to be bound by it under Paragraph 7.1. However, the probable 'losing' Party is unlikely to be cooperative. Even if it did initially agree, on finding out the result it might then retrospectively claim that Paragraph 7.1 was not applicable because the decision was not valid as it was reached too late.)

It is stated that the Adjudicator must act as an Adjudicator and not as an expert or an Arbitrator in reaching the decision. The reference to 'expert' is presumably not to 'expert witness' but to 'expert determination'. The need to make specific reference to 'expert'(or to 'expert determination') is not understood, nor why this is expressly stated but not, say, conciliation or mediation. If it is necessary, a definition is required, as different ADR terms often mean different things to different persons and organisations.

It is possible that this sentence is an attempt to better define the function of an Adjudicator. The Adjudicator is permitted to rely on its own knowledge and experience, but adjudication case law requires the Adjudicator to inform the Parties in the interests of natural justice when it is doing so, to give the Parties an opportunity to respond. This probably places the Adjudicator somewhere between an expert determiner, who might not have a duty to provide such disclosures, and an Arbitrator, whose own knowledge and experience will mainly be used to better understand the evidence provided by the Parties.

There will, inevitably, be adjudications that will tend more toward arbitration than expert determination, and vice versa. For example, it would be accepted generally that an expert determination comes about when two Parties agree to appoint a third party to use its knowledge and experience in order to provide a final and binding decision on a particular issue. Expert determinations are usually private contracts not susceptible to judicial review of matters such as not following the Rules of Natural Justice, but the expert does have a duty to observe procedural fairness as dictated by any agreed terms of reference, procedures or adopted code of conduct.

Agreed terms of reference might include that the expert can have separate private conversations with individual Parties; be inquisitorial; investigate and perhaps not be obliged to give the results of those investigations to the Parties before making a decision; provide a decision binding on the Parties; award interest; proceed even if one of the Parties at some stage refuses to cooperate; and be indemnified by the Parties. Terms of reference may also include that liability for the payment of fees will be joint and several, that there can be a lien on the decision and so on.

There thus appear to be marked similarities between an expert determination and an adjudication, particularly if in the latter case the Parties agree both to the Adjudicator and to the adjudication decision being final and conclusive. Indeed, the courts have suggested as much (*Bouygues* v *Dahl-Jensen* [1999 and 2000]).

It is therefore considered that the essential differences between expert determination and adjudication should have been clarified so that the drafters' causes for concern could be understood. On the face of it, this part of Paragraph 5.3 is not considered to be particularly helpful, and some might say it is unnecessary.

> *Provided that the Party who has made the referral may consent to allowing the Adjudicator to extend the period of 28 days by up to 14 days; and that by agreement between the Parties after the referral has been made a longer period than 28 days may be notified jointly by the Parties to the Adjudicator within which to reach his decision.*

Ignoring the redundant words shown underlined, the procedures for extending the 28 day timetable are clear. However, the Adjudicator must agree to any proposed timetable extension.

Paragraph 5.4 Reasons

> *The Adjudicator shall not be obliged to give reasons for his decision.*

This Paragraph reduces the likelihood of appeals on the Adjudicator's decision by not requiring the Adjudicator to provide reasons in its decision, unless it wants to.

Reasons are discussed in more detail in Chapter 13.

Paragraph 5.5 Impartiality, Adjudicator to set own procedures

> *In reaching his decision the Adjudicator shall act impartially, set his own procedure: and at his absolute discretion may take the initiative in ascertaining the facts and the law as he considers necessary in respect of the referral which may include the following:*
>
> 1. *using his own knowledge and/or experience;*
> 2. *subject to clause 30.9 opening up, reviewing and revising any certificate, opinion, decision, requirement or notice issued, given or made under the Contract as if no such certificate, opinion, decision, requirement or notice had been issued, given or made;*
> 3. *requiring from the Parties further information than that contained in the notice of referral and its accompanying documentation or in any written statement provided by the Parties including the results of any tests that have been made or of any opening up;*
> 4. *requiring the Parties to carry out tests or additional tests or to open up work or further open up work;*
> 5. *visiting the site of the Works or any workshop where work is being or has been prepared for this Contract;*

6. obtaining such information as he considers necessary from any employee or representative of the Parties provided that before obtaining information from an employee of a Party he has given prior notice to that Party;

7. obtaining from others such information and advice as he considers necessary on technical and on legal matters subject to giving prior notice to the Parties together with a statement or estimate of the cost involved;

8. having regard to any term of this contract relating to the payment of interest deciding the circumstances in which or the period for which a simple rate of interest shall be paid.

It will be noted in Paragraph 2 and the Definitions that an Adjudicator must be an individual and not an organisation. In addition, where an Adjudicator has been pre-named within the contract, page 55 of the Guidance Notes states that an Adjudicator should not normally be an employee or be otherwise engaged by either Party (e.g. any member of a Professional Team), unless both Parties agree.

To 'act impartially' is discussed in detail in Chapter 11 Procedural Fairness. Many of the listed matters are discussed in more detail in Chapter 12 Conduct of the Adjudication, which deals with practical procedures. There are some particular points that are appropriate to raise here:

An Adjudicator may open up, review and revise any certificate, decision, requirement or notice as if they had not been issued, given or made

This could apply to an Adjudicator opening up disputes that have been the subject of previous contractual adjudication decisions, including its own. Section 108 (3) of the Construction Act requires an adjudication decision to be binding until, *inter alia*, 'the dispute is finally determined ... by agreement'. Whilst there is nothing expressly stated in this part of Paragraph 5.5 about needing the agreement of the Parties, Paragraph 7.1 (see later in this text) expands on the requirements of Section 108 (3) by substituting for '... by agreement' the words ' ... or if agreed otherwise in writing between the Parties after the decision has been provided'. It is thus clear that the Parties can agree in writing that an Adjudicator can 'open up' etc. a previous adjudication decision, but that this must be a joint decision, not one taken by just one Party.

However, an Adjudicator cannot overturn a previous adjudication decision expressly stated by the Parties as being a final determination of a dispute. It will be noted that Section 108 (3) of the Construction Act expressly allows a decision to be a final determination if both Parties agree. Other decisions that cannot be overturned include those contractual procedures such as expert determination, conciliation, mediation and so on, which have been agreed to be final and binding.

As additional potential confusion, Paragraph 5.5 only applies to matters arising within the contract (such as, presumably, contractual conciliation requirements, e.g. ICE Procedure), so if expert determination, conciliation or any other final and conclusive agreement had been concluded privately outside the contract, those could not be opened up by an Adjudicator anyway.

In summary therefore, until decided by the courts where necessary, it is suggested that it would be prudent for an Adjudicator to assume that:

(a) It *cannot* 'open up, review' etc. 'any certificate, decision' etc. agreed by both Parties (either within or outside an underlying contract) to be final and conclusive, or expressly stated (within an underlying contract) to become final and conclusive by default if certain defined circumstances do not occur, notwithstanding that the adjudication procedures may be stated to take precedence over underlying contract clauses.
(b) It *can* open any other 'certificate, decision' etc. under this Paragraph, including previous adjudication decisions (as long as any other requirements are met, such as a need for the agreement of both Parties under Paragraph 7.1).

An Adjudicator may, after giving notice and a statement/estimate of costs, obtain technical and legal advice from others

Without the Parties having authority to prevent advice being obtained it is not clear what value a statement or estimate would be to the Parties, other than for information. In view of the limited time scale available for the decision, an Adjudicator will have proceeded forthwith with obtaining the required advice anyway. At most, if the costs are likely to be high, it might encourage the Parties to attempt to seek an alternative settlement, perhaps asking the Adjudicator to put a temporary hold on the adjudication process, as discussed in more detail in Chapter 12.

An Adjudicator can award simple interest

It is noted that Paragraph 30.1.1.1 in Amendment 18 states that simple interest shall be 5% over the Bank of England base rate pertaining when payment becomes overdue. Compound interest is therefore excluded by implication.

Paragraph 5.6 Validity of decision even when Parties do not cooperate

> *Any failure by either Party to enter into the JCT Adjudication Agreement or to comply with any requirement of the Adjudicator under* [Paragraph] *5.5 or with any provision in or requirement under clause 41A shall not invalidate the decision of the Adjudicator.*

Paragraph 5.7 Costs

> *The Parties shall meet their own costs of the Adjudication except that the Adjudicator may direct as to who should pay the cost of any test or opening up if required pursuant to* [Paragraph] *5.4.*

Although the Construction Act, does not mention costs, in this Paragraph there is an express requirement for the Parties to pay their own costs.

Paragraph 5.8 Disputes arising under Clause 8.4.4 of the Conditions of Contract

Where any dispute or difference arises under Clause 8.4.4 as to whether an instruction issued thereunder is reasonable in all the circumstances the following provisions shall apply:

1. *The Adjudicator to decide such dispute or difference shall (where practicable) be an individual with appropriate expertise and experience in the specialist area or discipline relevant to the instruction or issue in dispute.*
2. *Where the Adjudicator does not have the appropriate expertise and experience referred to in* [Paragraph] *5.8.1 above the Adjudicator shall appoint an independent expert with such relevant expertise and experience to advise and report in writing on whether or not any instruction issued under Clause 8.4.4. is reasonable in all the circumstances.*
3. *Where an expert has been appointed by the Adjudicator pursuant to* [Paragraph] *5.8.2 above the Parties shall be jointly and severally responsible for the expert's fees and expenses but, in his decision, the Adjudicator shall direct as to who should pay the fees and expenses of such expert or the proportion in which such fees and expenses are to be shared between the Parties.*
4. *Notwithstanding the provisions of* [Paragraph]*5.4 above, where an independent expert has been appointed by the Adjudicator pursuant to* [Paragraph] *5.8.2. above, copies of the Adjudicator's instructions to the expert and any written advice or reports received from such expert shall be supplied to the Parties as soon as practicable.*

The foregoing Paragraph is aimed at dealing with a particular aspect of the JCT Standard Form. As such, it is a clear enhancement to the minimum requirements of the Construction Act and does not result in the JCT Adjudication Procedure being non Construction Act-complaint. The requirements are straight forward and clear with only one possible exception: what is "appropriate expertise and experience" and who decides, particularly in borderline cases? As it is the Adjudicator that makes the expert appointment, it would seem to be a decision of the Adjudicator. This might be disputed by, say, a Responding Party that does not feel co-operative.

It would appear that in practice, with an appointed expert, the Adjudicator will usually be able to give the expert's views some legal status if that is needed, plus a means for rapid enforcement. It begs the question as to why the Parties do not consider authorising an expert determination or similar by a third party in the first place, with appropriate procedures to be followed and agreements put in place to achieve a similar effect to using an Adjudicator. As provided, an Adjudicator's role when an expert is used will surely in many cases be merely to rubber stamp the expert's report.

Paragraph 6.1 Adjudicator's fees and expenses

The Adjudicator in his decision shall state how payment of his fee and reasonable expenses is to be apportioned as between the Parties. In default of such statement the Parties shall bear the cost of the Adjudicator's fee and reasonable expenses in equal proportions.

The Adjudicator may therefore decide that one Party should pay more than half of its fees and expenses, because, for instance, that Party in the Adjudicator's view had been unreasonable, obstructive, and uncooperative during the adjudication process. The Adjudicator must be sure that it is allowed to do this, i.e. that it lies within its jurisdiction, otherwise a Party may consider that it has grounds for an appeal, on that basis or alternatively on the basis of partiality. However, as an Adjudicator does not have to give reasons, the Parties should tread carefully, as no reasons will mean it most unlikely that any appeal could be launched successfully against such a decision.

The Guidance Notes, page 45, paragraph (f), states, amongst other things, that:

The Parties may submit sealed letters to the Adjudicator, to be opened after he has reached his decision, if they wish him to consider particular matters before reaching his decision on apportionment of his fee and reasonable expenses under Paragraph 6.1 and costs under Paragraph 5.7.

An Adjudicator would normally have been intending to give its decision on all matters, including fees and costs, in accordance with the strict Construction Act timetable or any agreed amendment to that timetable. Under the Guidance Note, any time (even only a day or two) before that decision, one or more Parties can submit unexpected sealed letters of unlimited scope to the Adjudicator without copying them to the other Party. The Adjudicator is then obliged to consider the contents and possibly add to or amend part of its decision. If an Adjudicator has agreed to provide reasons, those reasons would now have to be reviewed in relation to how fees, expenses and costs have been apportioned. This could easily lead to delay in reaching a decision. There could also be claims for lack of partiality because the other Party could claim the Adjudicator had reached a decision on the basis of information not made available to it or to which it had had no opportunity to respond.

A solution, at least partially, to this particular issue (which in this text arises only within the JCT Procedure) would be for the Adjudicator to take control at an early stage and require any such submissions also to be provided openly to the other Party, by the same date that the last response was to be submitted to the Adjudicator, and that it should be in, or at least summarised in, no more than, say, two A4 sheets.

It is not clear exactly what the drafters had in mind in the Guidance Notes, but it is potentially highly significant to an Adjudicator. It may be that it is intended to cover offers made by one Party to the other which are rejected by the other Party but not exceeded by the Adjudicator in its award. That would mean that the adjudication had been needless and in court procedures that disentitles that Party to its costs. It may be that the drafters of the JCT Procedure had something similar in mind for adjudication. However, it must be noted that it only appears in the Guidance Notes and not in the Amendment and so its contractual status is arguable anyway.

Paragraph 6.2 Joint and several liability

The Parties shall be jointly and severally liable to the Adjudicator for his fee and for all expenses reasonably incurred by the Adjudicator pursuant to the Adjudication.

Paragraph 7.1 Binding decision

The decision of the Adjudicator shall be binding on the Parties until the dispute or difference is finally determined by arbitration or by legal proceedings or by an agreement in writing between the Parties made after the decision of the Adjudicator has been given.

One of the ways both Parties could agree to alter an adjudication decision is by having the same or another Adjudicator 'open up, review and revise' it in accordance with Paragraph 5.5 and agreeing to accept that second opinion. Associated issues have been discussed in detail within Paragraph 5.5. Alternatively, the Parties could both agree to abide by the result of another ADR method or to just agree a different decision between them.

Whether there is arbitration or litigation will depend upon which of Articles 7A or 7B is applicable. A footnote makes clear that subsequent arbitration or litigation under this Paragraph would not be an appeal against the adjudication decision, but a consideration of the dispute as if the earlier decision had never been made.

Paragraph 7.2 Compliance with decision

The Parties shall, without prejudice to their other rights under this Contract, comply with the decision of the Adjudicator; and the Employer and the Contractor shall ensure that the decision of the Adjudicator are given effect.

The reason for specific mention here of 'Employer and Contractor' instead of 'Parties' as used everywhere else, is unclear. The Employer and Contractor would normally be the Parties and thus the second part of the sentence would appear to be redundant. If not, then the reason is not understand, and should be explained (as well as, perhaps, what is meant or intended in practice by the word 'ensure').

Paragraph 7.3 Enforcement of decision

If either Party does not comply with the decision of the Adjudicator the other Party shall be entitled to take legal proceedings to secure such compliance pending any final determination of the referred dispute or difference pursuant to [Paragraph] *7.1.*

It will be noted that, unlike other adjudication procedures, there is no express mention of 'summary enforcement' or 'peremptory order', although, in practice, applications for summary judgment are probably the type of 'legal proceedings' that will be pursued in these circumstances. Enforcement is discussed in more detail in Chapter 15.

Paragraph 8 Immunity

> *The Adjudicator shall not be liable for anything done or omitted in the discharge or purported discharge of his functions as Adjudicator unless the act or omission is in bad faith and this protection from liability shall similarly extend to any employee or agent of the Adjudicator.*

This statement would only apply to the contracting, not to third, Parties. It would be better if the Parties provided an additional indemnity to the Adjudicator, similar to that in the ICE Procedure, e.g.:

> *The Parties shall save harmless and indemnify the Adjudicator and any employee or agent of the Adjudicator against all claims by third parties and in respect of this shall be jointly and severally liable.*

The Adjudicator could then pass all third party claims against it directly to the Parties to sort out. Otherwise, even if ultimately found not to be liable, an Adjudicator might have to spend considerable non-productive time significantly eroding limited profit margins in order to assist extricate itself from proceedings.

Even then, neither of the above would provide total Adjudicator immunity, even where an Adjudicator has not acted in bad faith.

This is firstly because if there is no specific reference to negligence, the most likely claim against an Adjudicator, there may be no immunity against a claim for it. It is considered that an additional reference to negligence should have been provided for absolute certainty of cover. The only other way to cover this elsewhere would have been in the Adjudicator's Agreement, but as discussed in Paragraph 2, amendments to this are, theoretically, not possible with pro forma non-alterable compulsory wording. This matter is discussed in more detail in Section 108 (4) of the Construction Act.

Secondly, the Parties (and the Adjudicator if not adequately indemnified against all possible actions by third parties in the Adjudication Procedure or in the Adjudicator's Agreement) might have potential liabilities under the Construction (Rights of Third Parties) Act 1999, e.g. this needs wording such as: 'For the avoidance of doubt in relation to the Construction (Rights of Third Parties) Act 1999, the Parties to this Adjudication do not intend that any of its terms or those in the Parties' agreement with the Adjudicator shall be enforceable against the Adjudicator (and the Parties) by any third party.'

Chapter 5

GC/Works Adjudication Procedure

General

The GC/Works/1 Without Quantities (1998) conditions of contract contain adjudication provisions that are intended to comply with the Construction Act. Those provisions are the basis of similar provisions in the other GC/Works/1, 2, 3 and 4 (1998) conditions of contract.

Adjudication requirements are described in Condition 59 of the GC/Works/1 conditions of contract [hereinafter the GC/Works Procedure], i.e. the adjudication provisions are incorporated directly in those conditions and not incorporated merely by reference.

Despite being based on the Construction Act requirements, the GC/Works Procedure is notably different from other standard ones, which, although varying from each other, all appear to have been approached on a similar basis.

The GC/Works Procedure is more complicated, particularly in three respects. Firstly, a Project Manager and Quantity Surveyor have a direct role in the adjudication, in addition to the Parties (although other professionals such as an architect, civil or structural engineer do not have a recognised individual role). Secondly, to determine the powers of an Adjudicator, reference must be made to the powers of an Arbitrator in the 1996 Arbitration Act and hence to published interpretations and clarifications. Thirdly, the total provisions are contained in several documents, e.g. Condition 59 (Adjudication), the Abstract of Particulars and Addendum, the Model Forms and Commentary (including Model Form 8 Adjudicator's Appointment, the Summary of Condition Changes and the Commentary itself), Condition 60 (Arbitration and choice of law), the 1996 Arbitration Act and associated commentaries on it.

It is considered that the result of these additional complications, intentionally or otherwise, is either likely to deter adjudications from taking place or else lead to the Referring Party looking to identify (or even deliberately creating – see later) non-compliances with the adjudication requirements of the Construction Act so that it can choose the more simple Scheme instead.

The following considers the detailed requirements. Condition 59 Paragraphs are quoted in full and given in italics. Other paragraphs are generally paraphrased and reference to original text will be needed to see the full wording in its correct context.

Condition 59 (Adjudication)

For simplification and ease of reference in the following, Condition 59 (1), 59 (2) etc. are referred to as Paragraph 1, 2 and so on. Also, in the commentary, the Employer and the Contractor are referred to as the Parties and 'disputes, differences or questions' are referred to as 'disputes'.

Paragraph 1 Notification of dispute, referral to Adjudicator

> *The Employer or the Contractor may at any time notify* [i.e. give notice to] *the other of* [their] *intention to refer a dispute, difference or question arising under, out of, or relating to, the Contract to adjudication. Within 7 Days of such notice* [of Adjudication], *the dispute, may by further notice* [of Referral] *be referred to the Adjudicator specified in the* [Condition 59] *Abstract of Particulars.*

It will be noted that disputes 'arising under, out of, or relating to' a contract is all-embracing, and is not restricted to breach of contract only. This issue is considered further in Chapter 2, under Section 108 (1).

As discussed under Section 108 (2)(b) in Chapter 2 and under the heading 'Urgency of appointment' in Chapter 9, it is not clear whether a referral within 7 days is a mandatory requirement of the Construction Act. Paragraph 1, as written, appears to believe that is not the case (and, that being so, there is no need to state what should happen if the 7 day timetable is not met). It is considered that this may be a future issue for the courts to clarify.

Paragraph 2 Notice of referral

> *The notice of referral shall set out the principal facts and arguments relating to the dispute. Copies of all relevant documents in the possession of the party giving the notice of referral shall be enclosed with the notice. A copy of the notice and enclosures shall at the same time* [as it is sent to the Adjudicator] *be sent by the party giving the notice to the PM* [the Project Manager], *the QS* [the project Quantity Surveyor] *and the other party.*

As described in the Commentary, page 100, this will usually mean that a Contractor will be sending the notice to the Employer and the Employer's PM and QS, as well as to the other Party and the Adjudicator.

Paragraph 3 Appointment of Adjudicator

> (*a*) *If the person named as Adjudicator in the Abstract of Particulars is unable to act, or is not or ceases to be independent of the Employer, the Contractor, the PM and the QS, he shall be substituted as provided in the* [Condition 59] *Abstract of Particulars.*
>
> (*b*) *It shall be a condition precedent to the appointment of an Adjudicator that he shall notify both parties that he will comply with this Condition* [59] *and its time limits.*

Paragraph 3(b) is incorporated into Model Form 8 Adjudicator's Appointment. If an alternative Agreement is used, or Model Form 8 is amended, the Paragraph 3(b) notification remains a condition precedent, i.e. it is mandatory and an express requirement for a valid appointment.

(c) *The Adjudicator, unless already appointed, shall be appointed within 7 Days of the giving of a notice of Intention to refer a dispute to adjudication* [i.e. the Notice of Adjudication] *under paragraph (1). The Employer and the Contractor shall jointly proceed to use all reasonable endeavours to complete the appointment of the Adjudicator and named substitute Adjudicator. If either or both such joint appointments has not been completed within 28 Days of the acceptance of the tender* [meaning?], *either the Employer or the Contractor alone may proceed to complete such appointments. If it becomes necessary to substitute as Adjudicator a person not named as Adjudicator or substitute Adjudicator in the Abstract of Particulars, the Employer and Contractor shall jointly proceed to use all reasonable endeavours to appoint the* [an agreed] *substitute Adjudicator. If such joint appointment has not been made within 28 Days of the selection of the substitute Adjudicator* [misplaced phrase?], *either the Employer or Contractor alone may proceed to make such appointment*

Without significant clarification (noting that the underlined phrases signifying suggested ommisions and comments in square brackets signifying suggested additions are the authors' attempts to make the paragraphs more comprehensible), the foregoing is unclear.

For all such appointments, the form of Adjudicator's appointment prescribed by the Contract shall be used, so far as is reasonably practicable. A copy of each such appointment shall be supplied to each party. No such appointment shall be amended or replaced without the consent of both parties.

Again, this part of the Paragraph can result in potential conflict. What does 'so far as is reasonably practicable' mean – does it refer to individual clauses or the whole Appointment? Who finally decides – does it have to be the Referring Party, the Responding Party or even the Adjudicator, or both Parties (as implied by the mention of those Parties in the last line)? When must this take place – pre or post contract award?

An Adjudicator will also have a view on what is reasonable or 'reasonably practicable'. Having a pro forma Adjudicator's Agreement will not prevent an Adjudicator from proposing what it considers to be reasonable additional terms and conditions, or even variations from what might be in the pro forma Agreement, adjudication procedures, or the Construction Act (e.g. the Parties can even agree to waive statutory requirements in individual circumstances as long as the waiver is clearly expressed). However, having made a proposal does not mean the same as having it accepted. At the commencement of an appointment, the Adjudicator could agree not to commence at all if its proposals are not met. If proposals are made during the adjudication and not agreed, the only options are for the Adjudicator to proceed anyway as if the proposals had not been made, or to resign.

For example, there is nothing mentioned in the pro forma Adjudicator's Agreement about providing the Adjudicator with an Advance Payment as security for some, or all, if its costs. This does not mean that the Adjudicator cannot ask for an Advance Payment or that the Parties cannot agree to provide one. Equally, they are under no obligation to do so. The issue is one between the Parties and the Adjudicator and if the Adjudicator considers the matter important enough it can decide not to proceed with the appointment.

Another example is that there is nothing mentioned in the pro forma Adjudicator's Agreement about paying the Adjudicator its outstanding fees and expenses before it releases its decision. If raised at the time of the proposed appointment, the default option for the Adjudicator is not to proceed with the appointment. If proposed by the Adjudicator during the adjudication and not accepted by the Parties, the Adjudicator has no option but to resign or continue, and, if it continues, to reach and release its decision within the stated Paragraph 5 timescale.

It is quite possible at this stage, with the likelihood of an Adjudicator not accepting an appointment or resigning, that the Parties themselves might reconsider what is 'reasonably practicable' to them and the successful resolution of the dispute.

It is suggested that in making proposals similar to the foregoing, the Adjudicator should make it clear, certainly to relatively 'unsophisticated' or legally unrepresented Parties, that they are under no obligation to agree to those proposals. Otherwise it is quite possible that complaints might be made, not least to the Adjudicator's appointing body, by a Party that feels it has been misled. The issue of obligatory pro forma Adjudicator agreements is discussed in more detail in Chapter 9 Adjudicator Appointments.

Paragraph 4 Response within 7 days

> *The PM, the QS and the other* [Responding] *party may submit representations to the Adjudicator not later than 7 Days from the receipt of the notice of referral.*

The Responding Party could be the Employer. Presumably any architect, Engineer, or other Employer's Consultant that is not the PM or QS, can contribute through the Employer. It will be noted that express provision for third parties to respond to an Adjudicator is most unusual in adjudication procedures.

If the Responding Party is the Contractor, this Paragraph provides the Employer's PM or QS with the opportunity to make further submissions on behalf of the Employer, but with no provision for a follow-up response by the responding Contractor.

The 7-day response time for a Responding Party will be extremely (and probably unnecessarily) short for a reasonable review and response to a referral in many disputes. It is unfortunate that this was not made more flexible. (It will be noted that any agreed timetable extension under Paragraph 5 (following) would apply to the time for a decision by the Adjudicator and not to the time for the response.)

It does not say so, but it is assumed the date of receipt is the date the Notice of Referral is actually received by the Adjudicator. There is no stated fall-back position if the Responding Party, the PM or QS receive their copies of the referral, accidentally or otherwise, a day or so after the Adjudicator, thus eroding further the time available for response.

Paragraph 5 Timetable for decision

> *The Adjudicator shall notify his decision to the PM, the QS, the Employer and the Contractor not earlier than 10 and not later than 28 Days from receipt of the notice of referral, or such longer period as is agreed by the Employer and the Contractor after the dispute has been referred.*
>
> *The Adjudicator may extend the period of 28 Days by up to 14 Days, with the consent of the party by whom the dispute was referred.*
>
> *The Adjudicator's decision shall nevertheless be valid if issued after the time allowed.*

To interpret 'notify' as 'advise the existence of' assists an Adjudicator requiring payment of fees and expenses before releasing the decision, as permitted by Section 56 of the Arbitration Act 1996 (see later in this chapter). This also makes sense of the second part of the the Paragraph, as well as being in accordance with Section 108 (2)(c) of the Construction Act which refers only to 'reaching' a decision within the required timetable.

If, alternatively, the interpretation of 'notify' is to 'provide' the decision, it is difficult to understand why there is a 28-day, or extended time limit if validity, according to the third part of the Paragraph remains for an undefined period thereafter if it is not met. The latter also provides a possible conflict with the strict timetable intention of Section 108 (2)(c) of the Construction Act, if 'reaching' is also to be taken as synonymous with 'provide', but issuing is permitted to be delayed by this Paragraph.

> *The Adjudicator's decision shall state how the cost of the Adjudicator's fee or salary (including overheads) shall be apportioned between the parties, and whether one party is to bear the whole or part of the reasonable legal and other costs and expenses of the other, relating to the adjudication.*

This Procedure, unlike others, expressly requires the Adjudicator to allocate costs between the Parties. Costs are discussed further in Chapter 13.

Paragraph 6 Adjudicator may take initiative

> *The Adjudicator may take the initiative in ascertaining the facts and the law, and the Employer and the Contractor shall enable him to do so.*
>
> *In coming to a decision the Adjudicator shall have regard to how far the parties have complied with any procedures in the Contract relevant to the matter in dispute and to what extent each of them has acted promptly, reasonably and in good faith.*

An Adjudicator must take into account how far a Referring Party has complied with, say, any notification and provision-of-details contract clause requirements in its decision. For example, if set-off is forbidden under a main contract, this should also be excluded in associated adjudications. However, in other contracts set-off may be allowed, but probably requires specific notices to be given and there might be a time limit for those notices. The use of the phrases 'still have regard to' and 'to what extent' are interesting in that it could be taken, for example, to imply that where there was a delay in issuing any appropriate notice, if in the Adjudicator's opinion that delay did not prejudice the other Party, it should have little or no effect.

An Adjudicator can usually only make a decision based on the facts and law (including the statutory requirements as stated for instance in the Construction Act and the Arbitration Act 1996). Powers for an Adjudicator to make a decision for other reasons can be agreed by the Parties in accordance with Section 46 of the Arbitration Act 1996 (see later in this chapter). Section 41 of the Arbitration Act also provides relevant powers to an Adjudicator in certain circumstances (see also later in this chapter). At the time of the decision the only action the Adjudicator can still take is in the apportionment of fees and costs within that decision (see Paragraph 5). An Adjudicator would be well advised to apply the provision of the second part of Paragraph 6 very carefully, particularly where reasons are to be given, otherwise there might be appeals on the basis of exceeding jurisdiction, partiality and so on.

> *The Adjudicator shall act independently and impartially, as an expert Adjudicator and not as an Arbitrator.*

It will be noted that there is a requirement to act 'independently' as required by the European Convention on Human Rights and some other contractual adjudication procedures. Impartiality is also mentioned in Model Form 8 Adjudicator's Appointment (see later) and in more detail in Chapter 11 Procedural Fairness.

The phrase that the Adjudicator shall not act as an 'expert or an Arbitrator', or similar, is used in some other adjudication procedures, where the former is taken to refer to 'expert determination'. In this instance, the phrase 'expert Adjudicator' has been used, and it is assumed that this is also intended to refer to expert determination. Without some confirmation and description or definition of what the latter is, with different ADR terms often meaning different things to different organisations, it is not clear what point the drafters are making.

It is possible that it is an attempt to better define the function of an Adjudicator. An Adjudicator is permitted to rely upon its own knowledge and experience, but adjudication case law requires the Adjudicator to inform the Parties, in the interests of natural justice, when it is doing so in order to give the Parties an opportunity to respond. This probably places an Adjudicator somewhere between an expert determiner (who might not have a duty to disclose such matters) and an Arbitrator (whose own knowledge and experience will mainly be used to better understand the evidence provided by the Parties).

There will, inevitably, be adjudications that tend towards arbitration or expert determination. Indeed, the courts have suggested as much (*Bouygues* v *Dahl-Jensen* [1999 and 2000]). Overall, on the face of it, this part of Paragraph 6 is not considered to be particularly helpful, and some might say it is unnecessary.

> *The Adjudicator shall have all the powers of an Arbitrator acting in accordance with Condition 60 (Arbitration and choice of law).*

Condition 60 powers are paraphrased later in this text. What is interesting is that many of the powers require the agreement of the Parties. It is just possible that a Referring Party might gain the option to use the Scheme by stating an intended refusal to agree to the use of one or more of the Condition 60 powers, where the lack of that power would result in the GC/Works adjudication procedure becoming non-Construction Act-compliant. An Adjudicator also has:

> *the fullest possible powers to assess and award damages and legal and other costs and expenses; and, in addition to, and notwithstanding the terms of, Condition 47 (Finance charges), to award interest. In particular, without limitation, the Adjudicator may award simple or compound interest from such dates, at such rates and with such rests as he considers meet the justice of the case –*
>
> *(a) on the whole or part of any amount awarded by him, in respect of any period up to the date of the award;*
>
> *(b) on the whole or part of any amount claimed in the adjudication proceedings and outstanding at the commencement of the adjudication proceedings but paid before the award was made, in respect of any period up to the date of payment;*
>
> *and may award such interest from the date of the award (or any later date) until payment, on the outstanding amount of any award (including any award of interest and any award of damages and legal and other costs and expenses).*

The Adjudicator can thus award simple or compound interest.

Paragraph 7 Binding decision

> *Subject to the proviso to Condition 60 (1) (Arbitration and choice of law), the decision of the Adjudicator is binding until the dispute is finally determined by legal proceedings, by arbitration (if the Contract provides for arbitration or the parties otherwise agree to arbitration), or by agreement: and the parties do not agree to accept the decision of the Adjudicator as finally determining the dispute.*

Section 108 (3) of the Construction Act expressly permits the Parties to accept an original Adjudicator's decision as a final determination. Where an adjudication decision has not been agreed as a final determination by the Parties, the Parties could later both agree to alter an adjudication decision by having the same or another Adjudicator 'open up, review and revise' it and agreeing to accept that second opinion. Alternatively the Parties could both agree to abide by the result of another ADR method or to just agree a different decision between them.

The Commentary on Condition 60, Paragraph 1 (see later) is also directly relevant to this issue. Essentially, unless a Party serves a notice of arbitration within 56 days of notification of the adjudication decision, that decision becomes 'unchallengeable' by arbitration. It is thus equivalent to Clause 66 (9)(b) of the ICE Conditions of Contract, where the end consequence is a 'final and binding' decision.

Paragraph 8 Adjudicator's powers to overrule

In addition to his other Powers, the Adjudicator shall have power to vary or overrule any decision previously made under the Contract by the Employer, the PM or the QS, other than decisions in respect of the following matters –

(a) decisions by or on behalf of the Employer under Condition 26 (Site admittance);

(b) decisions by or on behalf of the Employer under Condition 27 (Passes) (if applicable);

(c) provided that the circumstances mentioned in Condition 56 (1)(a) or (b) (Determination by Employer) have arisen, and have not been waived by the Employer, decisions of the Employer to give notice under Condition 56 (1)(a), or to give notice of determination under Condition 56 (1);

(d) decisions or deemed decisions of the Employer to determine the Contract under Condition 56 (8) (Determination by Employer);

(e) provided that the circumstances mentioned in Condition 58A (1) (Determination following suspension of Works) have arisen, and have not been waived by the Employer, decisions of the Employer to give notice of determination under Condition 58A (1); and

(f) decisions of the Employer under Condition 61 (Assignment).

In relation to decisions in respect of those matters, the Contractors's [sic] only remedy against the Employer shall be financial compensation.

The Construction Act adjudication provisions are to apply to any contractual dispute, and this could mean other matters for which financial compensation alone is not considered adequate by the Referring Party, e.g. an Adjudicator might be approached to specifically decide on the validity of a decision, a stated objective of a Referring Party being to have the decision changed back again.

This Paragraph now states that certain decisions cannot be varied or overruled, although an Adjudicator can be appointed to decide what financial compensation is due instead. Page 99 of the Commentary explains that this is to reflect what will be the Employer's view, that in practice some decisions cannot be reversed (although, of course, a Referring Party may not necessarily agree).

Notwithstanding the practicality or otherwise of varying a decision (and as noted, this itself may be disputed), strictly speaking adjudication procedures that do not permit an Adjudicator to decide a disputed decision in a 'construction contract' appear to be non-compliant with the Construction Act. Resolution would appear to be a matter for the courts, whether on a general basis or with reference to a particular issue arising under Paragraph 8's exceptions. If there is non-compliance with the Construction Act, the result is that these complex GC/Works adjudication procedures can be ignored and the Scheme, may be used wholly or partly (see commentary under Section 108 (5) in Chapter 2), instead.

Paragraph 9 Compliance with decision, enforcement

Notwithstanding Condition 60 (Arbitration and choice of law), the Employer and the Contractor shall comply forthwith with any decision of the Adjudicator; and shall submit to summary judgement and enforcement in respect of all such decisions.

The courts have stated (*Macob* v *Morrison* [1999]) that the usual procedure following an adjudication decision, *particularly in money cases*, would be for a Party to issue court proceedings claiming the amount due, followed by an application to the courts for summary judgment, all in accordance with normal debt recovery procedures. This procedure could also be enhanced by a Party requesting a court to reduce the normal timetable for debt recovery, e.g. the intimation period (the period in which a defendant must reply if it wishes to defend an action) (*Outwing* v *Randell* [1999]).

Summary enforcement is discussed in more detail in Chapter 13 The Decision.

Paragraph 10 Reasons

If requested by one of the parties to the dispute, the Adjudicator shall provide reasons for his decision. Such requests shall only be made within 14 Days of the decision being notified ['issued' or 'advised as available'?] *to the requesting party.*

On the face of this Paragraph a request for reasons could be made 14 days either side of notification of the decision. However, the first sentence states that reasons and decisions must go together if requested by either of the Parties. It is therefore considered that what is intended is that a request for reasons should be made no later than 14 days before the last day for notifying the decision. This would then provide an Adjudicator with reasonable notice of what was required and give it the opportunity to present a full decision with reasons including apportionment of fees and costs as required by Paragraph 5.

A sensible way forward would be for the Adjudicator to ask as early as possible before completing its decision, whether either or both Parties want reasons, stating that the Party(s) requesting them would be charged for the relevant costs. The 14-day limit would still apply.

Reasons are considered in more detail in Chapter 13.

Paragraph 11 Immunity

> *The Adjudicator is not liable for anything done or omitted in the discharge or purported discharge of his functions as Adjudicator, unless the act or omission is in bad faith. Any employee or agent of the Adjudicator is similarly protected from liability.*

This statement would only apply to the contracting, not to third, parties. It would be better if the Parties provided an additional indemnity to the Adjudicator, similar to that in the ICE Procedure, e.g.:

> *The Parties shall save harmless and indemnify the Adjudicator and any employee or agent of the Adjudicator against all claims by third parties and in respect of this shall be jointly and severally liable.*

The Adjudicator could then pass all third party claims against it directly to the Parties to sort out. Otherwise, even if ultimately found not to be liable, an Adjudicator might have to spend considerable non-productive time significantly eroding limited profit margins in order to assist extricate itself from proceedings.

Even then, neither of the above would provide total Adjudicator immunity, even where an Adjudicator has not acted in bad faith.

This is firstly because if there is no specific reference to negligence, the most likely claim against an Adjudicator, there may be no immunity against a claim for it. It is considered that an additional reference to negligence should have been provided for absolute certainty of cover. The only other way to cover this elsewhere would now be in the Adjudicator's Agreement. Fortunately, there is no absolute obligation to use Model Form 8 Adjudicator's Appointment (see later), i.e. the Introduction to the Model Forms and Commentary says that the Forms 'may be used'. Nor does Paragraph 3 of Condition 59 create such an obligation. There is thus scope for amendment. This matter is also discussed in more detail under Section 108 (4) in Chapter 2.

Secondly, the Parties (and the Adjudicator if not adequately indemnified against all possible actions by third parties in the Adjudication Procedure or in the Adjudicator's Agreement) might have potential liabilities under the Construction (Rights of Third Parties) Act 1999. this needs wording such as: 'For the avoidance of doubt in relation to the Construction (Rights of Third Parties) Act 1999, the Parties to this Adjudication do not intend that any of its terms or those in the Parties' agreement with the Adjudicator shall be enforceable against the Adjudicator (and the Parties) by any third Party.'

Abstract of Particulars and Addendum Condition 59 (Adjudication)

The 'Abstract of Particulars and Addendum' forming part of GC/Works/1(1998) also contains adjudication requirements relating to Condition 59:

1. A first choice Adjudicator can be named under this Condition.
2. If the first Adjudicator is unwilling or unable to act, or is not independent of the Parties, the PM or QS, a replacement Adjudicator can also be named under this Condition.
3. If the second Adjudicator is unwilling or unable to act, or is not independent of the Parties, the PM or QS, a replacement Adjudicator can be agreed between the Parties.
4. If the Parties cannot agree, either Party can approach the President or Vice-President of the Chartered Institute of Arbitrators to choose an Adjudicator. A similar procedure applies to equivalent Scottish contracts.
5. A model form, Model Form 8 Adjudicator's Appointment, is provided.
6. A footnote states that the Employer must name the same Adjudicators in all its related project contracts, e.g. with Contractors or professionals. (It will be noted, however, that different Adjudicators might still be involved in related contracts, e.g. if the first Adjudicator is involved in an adjudication between the Employer and Contractor, but is unable (say through potential overcommitment) to act in one between the Employer and the QS.)

Model Forms and Commentary. Summary of Condition 59 Changes

The Summary confirms, *inter alia*, that an Adjudicator has the power to award costs, that a decision is binding until varied or overruled by arbitration or the courts and must be complied with straight away, that an Adjudicator has express powers to vary or overrule decisions by or on behalf of an Employer (although that might be limited to financial compensation in some cases) and that Adjudicators are jointly appointed.

Model Forms and Commentary. Commentary on Condition 59

The Commentary repeats much that is in Condition 59. Particular points include:

1. All disputes must be adjudicated by a 'fully independent' person. (See Model Form 8 Adjudicator's Appointment, Paragraph 2 (described later in this chapter) for further explanation of what this is intended to include.)
2. A named Adjudicator and substitute Adjudicator must, if possible, be appointed before or at the time of the contract.
3. The same Adjudicators should be named in all associated contracts and sub-contracts; in the latter case the conditions of contract should require the main contractor to do so in its sub-contracts. This should assist prevent different decisions by different Adjudicators on essentially the same issues.

4. Although all Employer's decisions can be adjudicated, some in practice may not be reversible, but can be the subject of financial compensation instead. (This has been discussed in more detail earlier in this chapter in Paragraph 8 of Condition 59.)
5. An Adjudicator has full powers to correct errors and mistakes made by either side which have led to disputes.
6. An Adjudicator can award simple or compound interest by having the powers of an Arbitrator given in the 1996 Arbitration Act.
7. A Notice of Referral must specify the matter in dispute, the principal facts and arguments and copies of all relevant documents in the possession of the Referring Party. This may require 'skilful legal drafting and assembly' and combine the equivalent arbitration steps of 'claimant's points of claim and discovery of documents'. (A Responding Party – more likely to be an Employer – is unlikely to have the time to put together an equivalently detailed response, but hopefully for both Parties this will be entirely unnecessary in many disputes about relatively simple technical, quantum or extension of time issues, which may need little, if any, input from lawyers.)
8. Extensive reasoning in a decision may not be possible, but a Party 'may seek reasons within 14 days'. (This issue has been discussed earlier in this chapter in Paragraph 10 of Condition 59.)
9. A 'sensible' Party is unlikely to refuse an Adjudicator's request for an extension of time because it may as a consequence receive an adverse decision, e.g.:
   (a) If a Referring Party refuses to agree an extension, this provides an Adjudicator with a reason, as it states, to 'throw out the referring Party's claim' on the basis of there being insufficient time to consider the Responding Party's case.
   (b) If a Responding Party refuses to agree to an extension, particularly if that Party's case looks weak anyway, the claim might be allowed.

Point 9 seems somewhat draconian, but reference should be made to Section 41 of the Arbitration Act 1996 (see later in this chapter) which provides powers to an Arbitrator (and hence in this case to the Adjudicator) to do similar things.

However, dismissing or refusing claims should be done with care and provide no reason for an Adjudicator to forfeit its fees because it threw out a claim solely because it could not meet its pre-agreed 28-day timetable. By accepting an appointment, an Adjudicator has also accepted an obligation to meet the timetable come what may, even if that results in a rough and ready decision. If the Adjudicator felt the task had become impossible, a better alternative might be to threaten to and, if necessary, actually resign, with whatever fee reimbursement consequences follow from that.

The only reasonable sanction (and this is also mentioned in Section 41 in relation to non-compliance with an order) appears to be on the apportionment of costs and expenses. However, an Adjudicator must make its decision on the basis of facts, the law (including statutory requirements such as complying with the Construction and Arbitration Acts) and also in accordance with any additional agreement made under Section 46 of the Arbitration Act 1996 (see later). If there is evidence (and there could be sufficient evidence in a decision with reasons) of anything else being taken into account as well or instead, this might constitute grounds for an appeal because of exceeding jurisdiction, exhibiting partiality (e.g. can an Adjudicator award costs against one of the Parties just because it does not get with that Party?) and so on. There might also be PI insurance cover implications, as cover is for legal liability only.

10. Of relevance to considering the need for extensions to the adjudication timetable is that an Adjudicator can issue a valid decision even if it does not meet a pre-agreed timetable. (This has been discussed earlier within this text in Paragraph 5 of Condition 59.)
11. Also of relevance, an Adjudicator has full power to award and apportion costs and expenses, which can be used to 'punish unmeritorious' conduct. This should be done with care for the reasons given within point 9 above.
12. A 'successful' Party can immediately try to enforce a decision by summary judgment and enforcement through the courts.

Point 12 will probably be successful if there is no triable defence. The courts are actively supporting such procedures (see Selected Adjudication Summaries).

Model Forms and Commentary. Supporting documentation

The Model Forms ... may be used ...

There is no absolute obligation to use the Model Forms.

Model Form 8 Adjudicator's Appointment

The form of Adjudicator's Appointment includes the following Paragraphs.

Paragraph 2

An Adjudicator confirms that it is 'independent' of the Parties, PM and QS, agrees to use reasonable endeavours to remain independent, and to notify the Parties (but not necessarily the PM or QS) promptly if that is, or could, change. The Adjudicator also confirms that it will act impartially.

Paragraph 4

An Adjudicator can take independent reasonable legal or other professional advice and recover the reasonable costs of that advice as expenses.

Paragraph 5

An Adjudicator (and any advisors and assistants to the Adjudicator) must comply with the Official Secrets Act, Section 11 of the Atomic

> Energy Act (if relevant) and treat all information as confidential. The Adjudicator must take 'all reasonable steps' to ensure its advisors and assistants do, in fact, comply.

How the latter might be done is not stated. It is assumed that a written statement of proposed compliance by those persons in a form of words provided by the Employer would be a minimum reasonable step.

This Paragraph is particularly interesting because, unlike other adjudication procedures, it recognises that an Adjudicator, although acting as an individual, might have the assistance of others. This might be necessary where there is a large amount of data (perhaps some of it specialised and outside the immediate capability of the Adjudicator) to process and analyse in a limited timescale, which does not actually need to be undertaken by the Adjudicator personally (although of course the Adjudicator must take responsibility for decisions based on the results). The consequence should be a better decision and (as such assistance will probably be less expensive rates than that otherwise charged by the Adjudicator) a cheaper one.

Condition 60 (Arbitration and the choice of law)

In the following, some of the parts of this Condition relevant to adjudication are given.

Paragraph 1

> In addition to its other powers [assumed to be those arising under the 1996 Arbitration Act (see later)], the Arbitrator shall have the fullest possible powers to:
>
> (a) rectify the contract;
> (b) order inspections, measurements and valuations;
> (c) vary or overrule any decision previously made under the contract by the Employer, the PM, the QS or an Adjudicator, provided that the Contractor's only remedy against the Employer in relation to decisions in respect of the matters listed in Paragraph 8 of Condition 59 shall be financial compensation;
> (d) order consolidation of the proceedings with other proceedings; and/or that concurrent hearings shall be held; and to make such orders and directions relating to such consolidation and hearings as it thinks fit; and
> (e) make orders, directions and awards in the same way as if all the procedures of the High Court of Justice in England (if the proper law of the contract is English law), the High Court of Justice in Northern Ireland (if the proper law of the contract is Northern Ireland law) or the Court of Session in Scotland (if the proper law of the contract is Scots law), including, without limitation, as to joining one or more defendants or defenders or joining co-defendants or co-defenders or third parties, were available to the Parties and to the Arbitrator.

Also in Paragraph 1:

> Where any dispute, difference or question has been referred to an Adjudicator under Condition 59 (Adjudication), and the Adjudicator has issued its decision, a Party shall not be entitled to refer such dispute, difference or question to arbitration, and the Adjudicator's decision thereon shall become unchallengeable, unless that Party serves a notice to refer to arbitration within 56 days of receipt of notification [assumed to be of the details, not the availability] of the Adjudicator's decision. This shall apply whether or not an Adjudicator has notified its decision within the agreed time limit.

This gives the latest time for a notice to refer to arbitration.

Paragraph 2

> Unless the Parties otherwise agree, no reference to an Arbitrator shall be made in respect of a dispute which has been referred to an Adjudicator until the notification by the Adjudicator of its decision, or the expiry of 28 days from the receipt by the Adjudicator of a Notice of Referral, or such longer time as is agreed for the Adjudicator's decision, whichever is the earlier.

This gives the earliest time for a notice to refer to arbitration.

Model Forms and Commentary. Summary of Condition 60 Changes

1. An Arbitrator has express additional powers to rectify a contract, order inspections, measurements and valuations, and to vary or overrule decisions.
2. Adjudication decisions cannot be challenged by arbitration unless a Notice of Arbitration is given within 56 days.
3. Adjudication decision enforcement is excluded from arbitration (i.e. it has to be through the courts).

Model Forms and Commentary. Commentary on Condition 60

The Commentary inevitably repeats much that is in Condition 60. Particular points include:

1. All disputes are referable to an Arbitrator, except those concerning the enforceability of an Adjudicator's decision which must be by the courts.
2. The statement that 'There are no longer any "final and conclusive" decisions that cannot be adjudicated or arbitrated'.

Point 2 must be considered to be somewhat controversial. It could apply to an Adjudicator opening up previous Adjudication decisions of firstly, others and secondly, its own decisions. However, Section 108 (3) of the

Construction Act requires an adjudication decision to be binding until, *inter alia*, 'the dispute is finally determined ... by agreement'. Therefore, for one Party to instruct an Adjudicator to review one of its own earlier decisions without having the agreement of the other Party would be in conflict with the 'by agreement' requirement of the Construction Act, although an instruction agreed by both Parties, would, of course, comply. The statement would, in theory anyway, appear to apply even to adjudication decisions that the Parties had previously decided would be (or by default had become) final and conclusive. This cannot be correct! In addition, Section 108 (3) of the Construction Act expressly permits the Parties to accept an original Adjudicator's decision as a final determination. To overturn the conclusive intention of this would be directly contrary to the Construction Act.

On the face of it, an Adjudicator using these Procedures could also open up any other non-adjudication decision, including those decided by any previous ADR procedure, such as an expert determination, conciliation or discussion, agreed by the Parties or stated in a contract to be, or become, final and binding.

There is a close analogy between the Parties agreeing under Section 108 (3) that an adjudication decision could be a final and conclusive determination and the Parties agreeing that, say, the conclusions of a conciliation should be also be taken as a final determination. A court might see no practical difference regarding any conclusive agreements made within a contract. Nor should an Adjudicator.

If expert determination, conciliation or any other final and conclusive agreement had been concluded privately outside the contract, it is possible that the courts might decide that, as they themselves would not normally be able to reopen such matters (as discussed in more detail in Chapter 14 Miscellaneous Issues), then they should not be reopened by an Adjudicator either.

In summary, therefore, until decided by the courts where necessary, it is suggested that it would be prudent for an Adjudicator to assume that it *cannot* 'open up, review' etc. 'any certificate, decision' etc. agreed by both Parties (either within or outside an underlying contract) to be final and conclusive, or expressly stated (within an underlying contract) to become final and conclusive by default if certain defined circumstances do not occur.

3. An example of the latter given in the Commentary is that an Adjudicator's decision cannot be arbitrated unless a 'preliminary notice' of arbitration has been given within 56 days of that decision. It will be noted that the 'preliminary notice' is not the same as the 'referral to arbitration' and the former may be substantially in advance of the latter.
4. A dispute cannot be referred to arbitration if it is 'in due course' of adjudication.
5. A dispute cannot be referred to arbitration until after completion, alleged completion, abandonment or determination of a contract.

Arbitration Act 1996. Sections relating to an Arbitrator's powers

Under the arbitration provisions of Condition 60 an Arbitrator has certain powers in addition to those powers arising under the 1996 Arbitration Act. (Note that the reference is to 'powers' only, not to immunity provisions or anything else.) The following Sections describe many of the powers of an Arbitrator under that Act.

Some powers exist unless both Parties agree otherwise; others require the agreement of the Parties before they are even available to the Arbitrator. This is interesting because it enables a Referring Party to withhold agreement and thus might provide scope for claiming non-compliance with a requirement of the Construction Act, so enabling the statutory Scheme to be selected instead. There are numerous texts and commentaries available on the interpretation of the Arbitration Act, but, not least for space reasons, these are outside the scope of the present text.

Section 34

Subject to any alternative agreement by both Parties on 'any matter' (i.e. unless the Parties agree that an Arbitrator shall not do so), an Arbitrator can decide procedural and evidential matters, as long as it is done fairly and impartially avoiding unnecessary delay and expense, including:

1. When and where any part of the arbitration will take place.
2. The language to be used, plus the need for document translation.
3. Whether and when to have written submissions and responses.
4. Whether, when and which documents should be disclosed and produced.
5. Whether, when and what questions should be answered by the Parties.
6. Whether, when, what and how rules shall apply to submitted oral or written evidence. (Note that this can encompass a requirement for evidence to be given by affidavit.)
7. Whether and to what extent initiative shall be taken in determining facts and law.
8. Whether and to what extent there should be written or oral evidence or submissions.
9. The timetable for compliance by the Parties and any extension(s).

Section 35

Parties can give an Arbitrator powers to consolidate proceedings or concurrent hearings.

Section 37

Unless otherwise alternatively agreed by both Parties, an Arbitrator may appoint experts, legal advisors or assessors and allow them to attend the proceedings.

Section 38

Unless otherwise alternatively agreed by both Parties, the general powers of an Arbitrator include:

1. To require a claimant to provide security for the costs of the arbitration. (This also applies to a counter-claimant as Section 82 (1), (which deals with more minor definitions), includes a counter-claimant within its definition of a claimant.) It should be noted that there are several qualifying matters that need to be looked at *vis-à-vis* security for costs, which are outside the scope of the present text.
2. For any property owned or in possession of one of the Parties which is related to the proceedings, to order inspection, photographing, preservation, custody or detention by the Arbitrator, an expert or a Party, to order samples to be taken, observations to be made or experiments to be conducted.
3. To require evidence to be given on oath or affirmation and administer that oath or affirmation.
4. To require the preservation of evidence. (It will be noted that this does not equate to a 'Minerva' type injunction to restrain a Party from actually removing assets until an award is given, if it is not also a preservation of evidence issue.)

Section 39

The Parties can agree specifically (i.e. the powers do not exist otherwise) that an Arbitrator can, similarly to final awards:

1. Make provisional awards, including the payment of money or allocation of property.
2. Order a payment on account towards the costs of the arbitration.

The foregoing are subject to any necessary adjustment on final determination by the Arbitrator.

An Arbitrator cannot, however, grant summary judgment. If a Party does not comply with a provisional (as opposed to final) award, there might be difficulty in enforcement. One solution is to obtain a peremptory order from the Arbitrator under Section 41 and then, if necessary, a Party can apply to the courts for an order under Section 42.

Section 41

Unless otherwise alternatively agreed by the Parties, the following applies:

1. If an Arbitrator is satisfied that a claimant (or counter-claimant as described in Section 38) has been inordinate and inexcusably slow

(usually from the perspective of the other Party) in pursuing its claim and there is a substantial risk of unfairness or serious prejudice to the other Party, the Arbitrator can dismiss the claim (or counter-claim).
2. If a Party cannot give 'sufficient cause' for not attending a hearing, or provide requested written evidence or submissions, an Arbitrator can proceed in any case and provide an award. (The Party must therefore be given a reasonable opportunity to present an explanation and the other Party should also be allowed to comment.)
3. If a Party without showing 'sufficient cause' fails to comply with an order, an Arbitrator can order peremptory compliance by a certain time. (Once again, a Party must therefore be given a reasonable opportunity to explain non-compliance, and the other Party must also be allowed to comment. This will inevitably delay final compliance further.)
4. Where such a peremptory order relates to a client providing security for costs and it is not complied with, an Arbitrator can dismiss the claim.
5. If either Party does not comply with a peremptory order, an Arbitrator can state that no reliance will be placed on the allegation or material that gave rise to the order, draw adverse inferences, proceed to an award anyway and apportion costs accordingly.

Section 46

An Arbitrator must decide a dispute in accordance with the law or 'such other considerations' chosen by the Parties.

This permits the Parties to request, say, a commercial decision, but they should be aware that as insurers are only concerned with paying for legal liabilities, without prior agreement with insurers they may not have a valid insurance claim. The latter is considered in more detail in Chapter 16 Insurance Implications.

Section 47

Unless otherwise alternatively agreed by both Parties, an Arbitrator can:

1. Make more than one award at different times on different aspects of an issue.
2. Make an award on the whole claim or on part of a claim or cross-claim, but must make that clear in its award.

Section 48

Unless otherwise agreed alternatively by both Parties, an Arbitrator can order:

1. Payment in any currency.
2. A Party to do or not do something.

3. Specific performance.
4. The rectification, setting aside or cancellation of a deed or other document.

Section 49

The Parties can agree to permit an Arbitrator to award simple or compound interest on the whole or part of an award from dates and at a rate it considers equitable to the circumstances, for periods specified in the Section.

Section 50

If Parties settle whilst an arbitration is still proceeding, unless otherwise alternatively agreed by both Parties, an Arbitrator can:

1. Order the termination of proceedings.
2. Produce an agreed award.

Section 56

An Arbitrator can refuse to release an award until its fees and expenses have been paid. A Party may make a payment into court in order to have an award released.

This does not require the agreement of the Parties, but might require some reviewing of earlier clauses for consistency with regard to words such as 'notification' and similar, relating to the decision.

Section 57

The Parties can agree to provide powers to an Arbitrator to correct or make a further award.

Even without agreement of the Parties (although they should be allowed to make representations to the Arbitrator if they want to), an Arbitrator can within 28 days of the date of an award (or later if agreed with the Parties) correct clerical mistakes, accidental errors, omissions, provide clarifications, remove ambiguities and make an additional award on matters presented but not dealt with in an award.

Section 58

Although an award can still be challenged by 'arbitral process of appeal or review', it is final and binding, unless otherwise agreed between the Parties, i.e. the Arbitrator has the power to make a final and binding award.

Chapter 6

CIC Model Adjudication Procedure

General

The previous three chapters have considered the adjudication procedures included within standard conditions of contract, i.e. the ICE, JCT and GC/Works 'families' of contracts. The standard conditions of contract contain or refer to other adjudication clauses or provisions which must be read in conjunction with the corresponding adjudication procedures.

However, stand-alone adjudication procedures are also available which are not related to specific conditions of contract, but which can be incorporated within, for example a, 'bespoke' (or indeed any form of) contract. The one that will be considered in this chapter is that produced by the Construction Industry Council (CIC).

The CIC Model Adjudication Procedure: Second Edition [hereafter the CIC Procedure] is intended to comply with the Construction Act. It can be incorporated directly, or by reference, within conditions of contract. Most of the provisions are identical, or identical in practice, to those in the ICE Adjudication Procedure, not least because some of the persons concerned are common to both, but also because the Construction Act imposes the same core requirements. The comments made with respect to the ICE Procedure are, where relevant, repeated in this chapter so that it remains more self-contained.

The following considers the detailed requirements.

The CIC Procedure

Paragraph 1 Fair, rapid and inexpensive adjudication

> *The object of adjudication is to reach a fair, rapid and inexpensive decision upon a dispute arising under the Contract and this procedure shall be interpreted accordingly.*

Fairness is not a requirement of the Construction Act as already discussed in Chapter 2, e.g. the different time available to each Party to present its respective side to a dispute is clearly not fair. The courts on a Scheme adjudication (*Macob* v *Morrison* [1999]), where there was no stated express requirement for fairness, agreed that there was no such need. For contractual adjudication procedures such as the CIC Procedure, where there is an express requirement for fairness, arguments about a decision being invalid because of procedural unfairness would appear to be inevitable.

The CIC Procedure is consistent throughout with the use of the word 'under' and not 'in connection with' or similar. The issue of disputes arising 'under' the contract or otherwise is discussed in more detail under Paragraph 108 (1) in Chapter 2.

Paragraph 2 Acting impartially

The Adjudicator shall act impartially.

The key requirement for an Adjudicator is to 'act impartially'. It will be noted that there is no express reference to 'independence' as in the European Convention on Human Rights and some other contractual adjudication procedures. These issues are discussed in more detail in Chapter 11 Procedural Fairness.

There is no specific requirement for an Adjudicator to be 'a natural person acting in a personal capacity' [The Scheme – see later] or 'a named individual' [The ICE Procedure], or similar. The CIC Procedure perhaps considers that it should go without saying that an Adjudicator (like a judge or Arbitrator) will be an individual. There is therefore no restriction on the Adjudicator being a self-employed individual or a consultant, or a named director, partner or individual working for a larger organisation. Indeed, in the latter case it is often the situation that a construction Arbitrator or expert is a 'named' person making personal decisions or providing personal opinions, whilst still being engaged by that organisation. Notwithstanding the foregoing, it will be noted that as written, an organisation (e.g. including a university department or even a set of chambers) can be the Adjudicator.

The pre-contract award selection of an agreed list of Adjudicators is not precluded and has several practical advantages to the Parties (e.g. range of expertise and availability of an immediate standby if one Adjudicator is not available), all at no significant administrative additional cost or time.

Paragraph 3 Adjudicator may take initiative

The Adjudicator may take the initiative in ascertaining the facts and the law. He may use his own knowledge and experience.

The adjudication shall be neither an arbitration nor an expert determination.

The need in the second sentence of this Paragraph to make specific reference to 'expert determination' is not understood, nor why expert determination is expressly mentioned but not, say, conciliation or mediation. It is suggested that a definition is needed, as different ADR terms often mean different things to different persons and organisations.

It is possible that this sentence is an attempt to better define the function of an Adjudicator. An Adjudicator is permitted to rely upon its own knowledge and experience, but adjudication case law requires the Adjudicator to inform the Parties, in the interests of natural justice, when it is doing so in order to give the Parties an opportunity to respond. This probably places an Adjudicator somewhere between an expert determiner (who might not have a duty to disclose such matters) and an Arbitrator (whose own knowledge and experience will mainly be used to better understand the evidence provided by the Parties). There will, inevitably, be adjudications that tend towards arbitration or expert determination.

For example, it would be accepted generally that an expert determination comes about when two Parties agree to appoint a third party to use its knowledge and experience in order to provide a final and binding decision on a particular issue. Expert determinations are usually private contracts not susceptible to judicial review of matters such as not following the Rules of Natural Justice, but the expert does have a duty to observe procedural fairness as dictated by any agreed terms of reference, procedures or adopted code of conduct. Agreed terms of reference might include that the expert can have separate private conversations with individual Parties; be inquisitorial; investigate and perhaps not be obliged to give the results of those investigations to the Parties before making a decision; provide a decision binding on the Parties; award interest; proceed even if one of the Parties at some stage refuses to cooperate; and be indemnified by the Parties. Terms of reference may also include that liability for the payment of fees will be joint and several, that there can be a lien on the decision and so on.

There can thus be marked similarities between an expert determination and an adjudication, particularly where in the latter case the Parties agree both to the Adjudicator and to the adjudication decision being final and conclusive. Indeed, the courts have suggested as much (*Bouygues* v *Dahl-Jensen* [1999 and 2000]).

It is therefore considered that the essential differences between expert determination and adjudication should have been clarified so that the drafters' causes for concern could be understood. Overall, on the face of it, this part of Paragraph 3 is not considered to be particularly helpful, and some might say it is unnecessary.

Paragraph 4 Binding decision

> *The Adjudicator's decision shall be binding until the dispute is finally determined by legal proceedings, by arbitration (if the contract provides for arbitration or the parties otherwise agree to arbitration) or by agreement*

A contractual adjudication clause such as this has still to be interpreted by the courts in relation to adjudication. By analogy with the judgments in the statutory Scheme adjudication cases, it is likely that by this Paragraph the Parties are agreeing that, even if the subject of a major error of fact or a minor procedural defect, any decision of the Adjudicator must be regarded as temporarily binding and observed (usually by payment of any award) pending final determination. Of course, in contractual adjudications, just as in statutory adjudications, a jurisdictional error, bad faith or the like is likely to remove the binding nature of any decision. Furthermore, at the same time as being 'temporarily' binding, the Parties can (either before or after any decision) agree that the decision of the Adjudicator shall be 'finally' binding.

One of the ways both Parties could 'agree otherwise' to alter an adjudication decision is by having the same or another Adjudicator 'open up, review and revise' it in accordance with Paragraphs 26 and 30 (see later) and agreeing to accept that second opinion. Interestingly, Paragraph 30 only needs the agreement of both Parties where it is proposed to have a different, not the same, Adjudicator, but there is no immediately obvious reason why both Parties cannot agree to use the same Adjudicator. Associated issues are discussed at length within Paragraph 26.

Alternatively the Parties could both 'agree otherwise' to abide by the result of another ADR method or to just agree a different decision between them.

Whether there is arbitration or litigation would depend upon the exact terms of the underlying contract and how, if at all, it may have been amended. It should be noted that subsequent arbitration or litigation under this Paragraph would not be an appeal against the decision, but a consideration of the dispute as if the earlier adjudication decision had never been made. It will be noted that there appears to be no bar to the decision of the Adjudicator being cited in evidence.

Paragraph 5 Implementation of decision without delay

The Parties shall implement the Adjudicator's decision without delay whether or not the dispute is to be referred to legal proceedings or arbitration.

A Party cannot delay complying with a decision until it is sorted out by arbitration or by the courts. The courts (e.g. *Macob* v *Morrison* [1999] and *Outwing* v *Randell* [1999]) have strongly supported this.

For an award of money it is assumed that 'without delay' must at the latest mean to comply with the appropriate payment provisions of the conditions of contract (which, of course, must comply with the relevant payment provisions of the Construction Act).

To enforce compliance, the other Party can make use of the enforcement provisions described in Paragraph 30 and its commentary.

Paragraph 6 Current edition

If this procedure is incorporated into the Contract by reference, the reference shall be deemed to be to the edition current at the date of the Notice [of Adjudication described in Paragraph 8 (see later)].

Paragraph 7 Precedence

If a conflict arises between this procedure and the Contract, unless the Contract provides otherwise, this procedure shall prevail.

Exact implications will therefore depend on the requirements of the underlying contract in each case.

Paragraph 8 Notice of Adjudication and contents

Either Party may give notice at any time of its intention to refer a dispute arising under the Contract to adjudication by giving a written Notice [of Adjudication] to the other Party. The Notice shall include a brief statement of the issue or issues which it is desired to refer and the redress sought. The referring Party shall send a copy of the Notice to any Adjudicator named in the Contract.

A development in statutory Scheme adjudication has been the importance which the courts have come to attach to clarifying what the dispute is and the extent to which it is correctly reflected in the Notice of Adjudication. How such judgments will apply to contractual adjudications is not yet clear. There might well be greater scope in a contractual adjudication (and in this case see Paragraph 20) for the Parties to agree between themselves and the Adjudicator any adjustment or enhancement of the description of the dispute being referred for adjudication. Notwithstanding any such agreement, it is suggested that even those involved in contractual adjudications should attempt to make the 'dispute' stated in the Notice of Adjudication as clear as possible.

Paragraph 9 Appointment and referral within 7 days

The object of the procedure in paragraphs 10 [to] *14 is to secure the appointment of the Adjudicator and referral of the dispute to him within 7 days of the giving of the Notice* [of Adjudication].

As discussed under Section 108 (2)(b) in Chapter 2 and under the heading 'Urgency of appointment' in Chapter 9, it is not clear whether the Adjudicator appointment and referral within 7 days is a mandatory requirement of the Construction Act. Paragraph 9, as written, does not clarify this either, (nor does it state procedures to follow if the 7 day timetable is not achievable). This issue may be for the courts to clarify.

Paragraph 10 Pre-named, stated body or CIC or agreed appointments

If an Adjudicator is named in the Contract, he shall within 2 days of receiving the Notice confirm his availability to act. If no Adjudicator is named, or if the named Adjudicator does not so confirm, the referring Party shall request the body stated in the Contract if any, or if none the Construction Industry Council, to nominate an Adjudicator within 5 days of receipt of the request.

The request shall be in writing, accompanied by a copy of the Notice and the appropriate fee.

Alternatively the Parties may, within 2 days of the giving of the Notice, appoint the Adjudicator by agreement.

Paragraph 11 Selection of replacement Adjudicator

If, for any reason, the Adjudicator is unable to act, or fails to reach his decision within the time required by this procedure, either Party may request the body stated in the Contract if any, or if none the Construction Industry Council, to nominate a replacement Adjudicator.

No such request may be made after the Adjudicator has notified the Parties that he has reached his decision.

This Paragraph does not, for some reason, expressly permit the Parties to simply agree a replacement Adjudicator between them. This appears an unnecessary limitation, particularly if that is how the original Adjudicator appointment was made under the alternative given in Paragraph 10.

It will be noted that Paragraph 25 also deals with replacing an Adjudicator, this time where there is a late notification of a decision.

Paragraph 12 Obligation to use CIC attached Agreement. Adjudicator entitlement to reasonable fees and expenses

Unless the Contract provides otherwise, the Adjudicator shall be appointed on the terms and conditions set out in the attached Agreement[.]

[The Adjudicator] *shall be entitled to a reasonable fee and expenses.*

The CIC terms and conditions for the appointment of an Adjudicator are clear but basic, and an Adjudicator might well prefer something different.

If the pro forma Adjudicator's Agreement is used, it requires signing by the Parties. No time limit is stated for signing, nor what an Adjudicator should do if that does not happen. However, it is suggested that any objection to an Adjudicator evidenced by a reluctance or refusal to sign the Agreement would have no effect, as described in following Paragraph 13. Both of these issues are discussed in Chapter 9 Adjudicator Appointments. If pre-named Adjudicators have been agreed, the foregoing might not be a problem if (similar in principle to, say, Forms of Bonds incorporated into construction contracts) it has been made a requirement of the Parties to sign any pre-named Adjudicator's Agreements as a prerequisite of the construction contract becoming valid.

Having a pro forma Adjudicator's Agreement does not prevent an Adjudicator from proposing additional terms and conditions, or even variations from what might be in the pro forma Agreement, adjudication procedures or the Construction Act (e.g. the Parties can even agree to waive statutory requirements in individual circumstances as long as the waiver is clearly expressed). However, having made a proposal does not mean the same as having it accepted. At the commencement of a referral, the Adjudicator could decide not to commence at all if its proposals are not met. If proposals are made during the adjudication and not met, the only options are for the Adjudicator to proceed anyway as if the proposal was not made, or to resign.

For example, there is nothing mentioned in the pro forma Adjudicator's Agreement about providing the Adjudicator with an Advance Payment as security for some, or all, of its costs. This does not mean that the Adjudicator cannot ask for an Advance Payment or that the Parties cannot agree to provide one. Equally, they are under no obligation to do so. The issue is one between the Parties and the Adjudicator and if the Adjudicator considers the matters important enough it can decide not to proceed with the appointment.

It is suggested that in making any proposals, the Adjudicator should make it clear, certainly to relatively 'unsophisticated' or legally unrepresented Parties, that they are under no obligation to agree to them. Otherwise it is quite possible that complaints might be made, not least to the Adjudicator's appointing body, if a Party feels it has been misled.

Paragraph 12 permits alternatives to the CIC Agreement if allowed for in the underlying contract.

Paragraph 13 Objection to Adjudicator to have no effect

If a Party objects to the appointment of a particular person as Adjudicator, that objection shall not invalidate the Adjudicator's appointment or any decision he may reach.

Presumably, if both Parties object there could be problems in the Adjudicator trying to continue, even though in theory the Adjudicator could do so if it has been validly appointed. Under Paragraph 29 the Referring Party is liable for outstanding fees and expenses if an adjudication does not proceed.

Paragraph 14 Statement of case

The referring Party shall send to the Adjudicator within 7 days of the giving of the Notice (or as soon thereafter as the Adjudicator is appointed) and at the same time copy to the other Party, a statement of its case including a copy of the Notice, the Contract, details of the circumstances giving rise to 'the dispute', the reasons why it is entitled to the redress sought, and the evidence upon which it relies. The statement of case shall be confined to the issues raised in the Notice.

This single Paragraph contains three different issues and would benefit from being rewritten and expressed in that way (underlining indicates words that could be omitted), i.e.:

The referring Party shall send to the Adjudicator w[W]*ithin 7 days of the giving of the Notice* [of Adjudication], *(or as soon thereafter as the Adjudicator is appointed)* [the referring Party shall send to the Adjudicator,] *and at the same time copy to the other Party, a statement of its case*[.] *including*

[The statement of case shall include] *a copy of the Notice* [of Adjudication], *the Contract, details of the circumstances giving rise to the dispute, the reasons why it is entitled to the redress sought, and the evidence upon which it relies.*

The statement of case which shall be confined to the issues raised in the Notice [of Adjudication].

Once a Notice of Adjudication has been issued (a document which an Adjudicator may have relied upon when accepting an appointment) the issues in the dispute should not be added to (except under Paragraph 20 (see later)).

There are a number of potential problems. For example, it is assumed that the 'giving' of the Notice refers to the 'date of receipt by the Adjudicator', but this could have been made clearer as it could be disputed. Another possible problem is what happens if an Adjudicator receives the documents promptly but, intentionally or otherwise, there is a delay in the other Party receiving them? Various issues of this type are discussed in more detail in Chapter 12 Conduct of the Adjudication, which deals with some practical procedures.

Paragraph 15 Date of referral

The date of referral shall be the date on which the Adjudicator receives this statement of case.

No matter how late an Adjudicator receives the Referring Party's full statement of case documents, the 28-day, or any extended, timetable does not commence until it does.

Paragraph 16 Decision in 28 days, or longer as agreed

The Adjudicator shall reach his decision within 28 days of the date of referral, or such longer period as is agreed by the Parties after the dispute has been referred. The Adjudicator may extend the period of 28 days by up to 14 days with the consent of the referring Party.

This conforms to Section 108 (2)(c) of the Construction Act. It will be noted firstly, that the Adjudicator must agree to any extension and secondly, that under Paragraph 24 the Adjudicator has merely to 'reach' (and probably notify the availability of) its decision in the required timescale (e.g. if its actual handing over depends on payment of an Adjudicator's fees, then that can be accommodated). It will be recognised that this interpretation of 'reach' may not accord with everyone's interpretation of Section 108 (2)(c) of the Construction Act, as some consider 'reach' to include an unconditional provision to the Parties as soon as possible after a decision has been reached.

Paragraph 17 Procedure to be at Adjudicator's complete discretion

The Adjudicator shall have complete discretion as to how to conduct the adjudication, and shall establish the procedure and timetable, subject to any limitation there may be in the Contract or the Act. He shall not be required to observe any rule of evidence, procedure or otherwise, of any court or tribunal. Without prejudice to the generality of these powers, he may:

(i) request a written defence, further argument or counter argument

(ii) request the production of documents or the attendance of people whom he considers could assist

(iii) visit the site

(iv) meet and question the Parties and their representatives

(v) meet the Parties separately

(vi) limit the length or time for submission of any statement, defence or argument

(vii) proceed with the adjudication and reach a decision even if a Party fails to comply with a request or direction of the Adjudicator

(viii) issue such further directions as he considers to be appropriate.

Paragraph 18 Compliance with requests or directions

The Parties shall comply with any request or direction of the Adjudicator in relation to the adjudication.

Requests and directions can be taken to be the same as instructions.

Paragraph 19 Legal and technical advice

The Adjudicator may obtain legal or technical advice, provided that he has notified the Parties of his intention first. He shall provide the Parties with copies of any written advice received.

This seems straightforward – if the Parties are expected to pay for something they should be told in advance, and should receive details of the advice obtained. This should apply to any advice that will be relied upon in the decision, even oral advice (although in practice a prudent Adjudicator should have confirmed the latter in writing anyway).

Paragraph 20 Matters to be dealt with by Adjudicator

The Adjudicator shall decide the matters set out in the Notice [of Adjudication], together with any other matters which the Parties and the Adjudicator agree shall be within the scope of the adjudication.

Paragraph 21 Application of law

The Adjudicator shall determine the rights and obligations of the Parties in accordance with the law of the Contract.

Paragraph 22 Joining another Party

Any Party may at any time ask that additional parties shall be joined in the adjudication. Joinder of additional parties shall be subject to the agreement of the Adjudicator and the existing and additional parties. An additional party shall have the same rights and obligations as the other [existing] Parties, unless otherwise agreed by the Adjudicator and the [existing and additional] Parties.

The joining of other Parties is discussed in some detail in Chapter 14 Miscellaneous Issues.

Paragraph 23 Resignation by Adjudicator

The Adjudicator may resign at any time on giving notice in writing to the Parties.

Nothing is stated about the length of notice to be given, but in practice that would probably not be an issue.

Nothing is stated about the circumstances or reasons for resignation that might give rise, or otherwise, to the payment or alternatively forfeiting, of outstanding fees and expenses. This would have been helpful. The issue is discussed in more detail in Chapter 9, Adjudicator Appointments.

Paragraph 24 Reaching decision. Withholding decision. Reasons

The Adjudicator shall reach his decision within the time limits in paragraph 16.

The Adjudicator may withhold delivery of his decision until his fees and expenses have been paid.

It is assumed that there is an implied obligation to notify the Parties when the decision has been reached. In practice, it is unlikely an Adjudicator would delay doing so. This is also discussed in the commentary after Paragraph 16.

He shall be required to give reasons unless both Parties agree at any time that he shall not be required to give reasons.

The obligation to give reasons in all circumstances, unless otherwise excused by both Parties, is unusual in most adjudication procedures. Reasons are discussed in more detail in Chapter 13 The Decision.

Paragraph 25 Decision remains effective unless dispute actually referred to a replacement Adjudicator. Adjudicator then forfeits entitlement to fees

If the Adjudicator fails to reach his decision within the time permitted by [paragraph 16 of] *this procedure, his decision shall nonetheless be effective if reached before the referral of the dispute to any replacement adjudicator under paragraph 11*[,] *but not otherwise.*

If he fails to reach such an effective decision, he shall not be entitled to any fees or expenses (save for the cost of any legal or technical advice[,] subject to the Parties having received such advice).

In other words, without reaching any agreement with the Parties for an extension, an Adjudicator might have a further minimum of 7 days (see Paragraph 9) in which to reach and notify the availability of its decision and not forfeit an entitlement to be paid its fees and expenses. This right of an Adjudicator to unilaterally have an extension to a specific timetable without reference to one or both Parties is a useful provision, as it is presumably more preferable (at least to the Referring Party) than having to start all over again. However, it does appear to be contrary to the timetable requirements of the Construction Act. It is certainly a useful practical fall-back position if an Adjudicator is unavoidably delayed for a short time, although it is possible, of course, that some Adjudicators will make regular use of this when organising their diaries.

Paragraph 26 Adjudicator may open up any certificate ...

The Adjudicator may open up, review and revise any certificate, decision, direction, instruction, notice, opinion, requirement or valuation made in relation to the Contract.

This could apply to an Adjudicator opening up previous Adjudication decisions of firstly, others (and this expressly requires agreement by both Parties in accordance with Paragraph 30) and secondly, its own decisions (about which Paragraph 30 is silent about a need for an agreement by both Parties). However, Section 108 (3) of the Construction Act requires an adjudication decision to be binding until, *inter alia*, 'the dispute is finally determined ... by agreement'. Therefore, for one Party to instruct an Adjudicator to review one of its own earlier decisions without having the agreement of the other Party would be in conflict with the 'by agreement' requirement of the Construction Act, although an instruction agreed by both Parties would comply.

However, an Adjudicator cannot overturn a previous adjudication decision expressly stated by the Parties as being a final determination of a dispute. It will be noted that Section 108 (3) of the Construction Act expressly allows a decision to be a final determination if both Parties agree. Other decisions that cannot be overturned include contractual procedures such as conciliation, mediation and so on, which have been agreed to be final and binding.

As an added potential confusion, Paragraph 26 only applies to matters arising 'in relation to' the contract anyway (e.g. such as any contractual conciliation requirements). If expert determination, conciliation or any other final and conclusive agreement had been concluded privately outside the contract, it is possible that the courts might decide that as they themselves would not normally be able to reopen such matters (as discussed in more detail in Chapter 14), then they should not be reopened by an Adjudicator either.

In summary therefore, until decided by the courts where necessary, it is suggested that it would be prudent for an Adjudicator to assume that:

(a) It *cannot* 'open up, review' etc. 'any certificate, decision' etc. agreed by both Parties (either within or outside an underlying contract) to be final and conclusive, or expressly stated within an underlying contract to become final and conclusive by default if certain defined circumstances do not occur, notwithstanding that the adjudication procedures may be stated to take precedence over underlying contract clauses.
(b) It *can* open any other 'certificate, decision' etc. under this Paragraph, including previous adjudication decisions (as long as any other requirements are met, such as a need for the agreement of both Parties).

Paragraph 27 Interest

The Adjudicator may in any decision direct the payment of such simple or compound interest from such dates, at such rates and with such rests, as he considers appropriate.

The Adjudicator can award simple or compound interest. Under common law, there is no power to award interest, so this has to be expressly given, as it is here. It will be noted that there is no stated requirement that there must be a claim for interest before an Adjudicator can award it. (See also under Interest in Chapter 13.)

Paragraph 28 Parties' costs

The Parties shall bear their own costs and expenses incurred in the adjudication.

Under this Paragraph there is an express requirement for the Parties to pay their own costs.

Paragraph 29 Liability for Adjudicator's fees and expenses

The Parties shall be jointly and severally liable for the Adjudicator's fees and expenses, including those of any legal or technical adviser appointed under paragraph 19, but the Adjudicator may direct a Party to pay all or part of the fees and expenses. If he makes no such direction, the Parties shall pay them in equal shares.

The Party requesting the adjudication shall be liable for the Adjudicator's fees and expenses if the adjudication does not proceed.

The Adjudicator may decide that one Party should pay more than half of its fees, because, for instance, that Party had, in the Adjudicator's view, been unreasonable, obstructive or uncooperative during the adjudication process. However, the Adjudicator must be sure that it is allowed to do this (i.e. that it lies within its jurisdiction and it has good grounds for doing so), otherwise a Party may consider appealing the decision on the basis of lack of jurisdiction, insufficient reason or, alternatively, on the basis of partiality. This is particularly so in these adjudication procedures, as an Adjudicator will normally be required to provide reasons for its decisions under Paragraph 24.

Paragraph 30 Summary enforcement. Decision not to be referred to another Adjudicator

The Parties shall be entitled to the redress set out in the decision and to seek summary enforcement, whether or not the dispute is to be finally determined by legal proceedings or arbitration.

The CIC Procedure does not state whether enforcement should be by the courts or arbitration, perhaps to avoid any discrepancy with what might be stated within the underlying construction contract. The contract might, for example, expressly require that enforcement of an adjudication decision should be through the contractual arbitration process. If there are no arbitration clauses (or later agreement to arbitrate), enforcement would have to be by the courts. The preferred means of enforcement will probably be by using the courts where possible, as they have consistently supported rapid enforcement (e.g. *Macob* v *Morrison* [1999]). In that case, the courts stated that the usual procedure following an adjudication decision, *particularly in money cases*, would be for a Party to issue court proceedings claiming the amount due, followed by an application to the courts for summary judgment, all in accordance with normal debt recovery procedures. This procedure could also be enhanced by a Party requesting a court to reduce the normal timetable for debt recovery, e.g. the intimation period (the period in which a defendant must reply if it wishes to defend an action) (*Outwing* v *Randell* [1999]). Summary enforcement is discussed in more detail in Chapter 15.

No issue decided by the Adjudicator may subsequently be referred for decision by another Adjudicator unless so agreed by the Parties.

This part of Paragraph 30 has already been noted under Paragraph 26.

Paragraph 31 Parties subsequent rights and obligations not affected

In the event that the dispute is referred to legal proceedings or arbitration, the Adjudicator's decision shall not inhibit the right of the court or Arbitrator to determine the Parties' rights or obligations as if no adjudication had taken place.

The purpose of this Paragraph is presumably to make clear that an Arbitrator or court will not, in the circumstances of the Paragraph, be assessing the correctness or otherwise of an adjudication decision, but will instead be considering the dispute anew.

It is noted that although the Adjudicator cannot be a witness in such proceedings (see Paragraph 32), there is no stated bar to disclosure of the Parties' submissions to the Adjudicator or of the Adjudicator's decision itself. Each Party can presumably present these to the relevant tribunal and this could lead to subsequent proceedings being significantly shortened.

The Paragraph would not apply anyway to any decision agreed as final and conclusive between the Parties as neither an Arbitrator nor the courts (as discussed in Chapter 14, Miscellaneous issues, as well as Paragraph 26 of this chapter) would reconsider such decisions.

Paragraph 32 Adjudicator not to be subsequent Arbitrator or witness

Unless the Parties agree, the Adjudicator shall not be appointed Arbitrator in any subsequent arbitration between the Parties under the Contract.

This is a total exclusion of the Adjudicator acting as an Arbitrator.

No Party may call the Adjudicator as a witness in any legal proceedings or arbitration concerning the subject matter of the adjudication.

This exclusion only applies to the adjudication the Adjudicator has been connected with.

The reason for the difference in the scope of exclusion between the two parts of the same Paragraph, if that is intended, is not clear.

Paragraph 33 Immunity of Adjudicator

The Adjudicator is not liable for anything done or omitted in the discharge or purported discharge of his functions as Adjudicator (whether in negligence or otherwise) unless the act or omission is in bad faith, and any employee or agent of the Adjudicator is similarly protected from liability.

This specific reference to negligence (the most likely claim against an Adjudicator) is necessary as discussed in more detail under Section 108 (4) in Chapter 2.

This statement would only apply to the contracting, not to third, Parties. It is suggested that it would be better if the Parties provided an additional indemnity to the Adjudicator similar to that in the ICE Procedure, e.g.:

The Parties shall save harmless and indemnify the Adjudicator and any employee or agent of the Adjudicator against all claims by third parties and in respect of this shall be jointly and severally liable.

The Adjudicator could then pass all third party claims against it directly to the Parties to sort out. Otherwise, even if ultimately not liable, an Adjudicator might have to spend considerable non-productive time significantly eroding limited profit margins in order to assist extricate itself from proceedings.

Paragraph 34 No duty of care to third parties

The Adjudicator is appointed to determine the dispute or disputes between the Parties and his decision may not be relied upon by third parties, to whom he shall owe no duty of care.

It is suggested that the Parties (and the Adjudicator if not adequately indemnified against all possible actions by third parties in the Adjudication Procedure or in the Adjudicator's Agreement) should seek advice on the effectiveness of the foregoing wording in reducing or eliminating any potential liabilities under the Construction (Rights of Third Parties) Act 1999, as the latter was finalised and came into force after the CIC Procedure wordings were published. It might now be prudent to use wording such as: 'For the avoidance of doubt in relation to the Construction (Rights of Third Parties) Act 1999, the Parties to this Adjudication do not intend that any of its terms or those in the Adjudicator's Agreement shall be enforceable against the Adjudicator (and the Parties) by any third Party.'

Paragraph 35 Law

This procedure shall be interpreted in accordance with the law of England and Wales.

Where appropriate substitute 'Scotland' for 'England and Wales' (and in Paragraph 30 delete the word 'Summary').

Chapter 7

The Scheme

General

As described under Section 108 (5) in Chapter 2 dealing with the Construction Act, if the adjudication provisions of any 'construction contract' as defined in the Construction Act do not meet each of its eight minimum adjudication requirements or the construction contract is silent regarding the rules to be applied, then, pursuant to Section 108 (5), the Scheme for Construction Contracts (England and Wales) Regulations 1998 [hereafter referred to as the Scheme] will automatically be available (i.e. can be implied into the construction contract) for the adjudication of disputes instead, if a Referring Party wishes to use it. Other points that are discussed in Section 108 (5) include how obligatory the use of the Scheme is, who has the authority to decide to use the Scheme and whether the whole or only part of the Scheme has to be substituted for non-Construction Act-compliant contractual adjudication provisions. The commentary on Section 108 (5) should be referred to for more detailed discussion on these issues.

It is reasonable to suppose that the Scheme is indicative of what was intended by the Construction Act.

The Scheme, in fact, went through several draft consultation phases and was widely amended before final publication. It attempts to strike a balance between the suggestions of various vested interests such as lawyers, Employers, professionals, Contractors and sub-Contractors, the vast majority of whom had no practical adjudication experience but often, it is suggested, had a strong view for or against adjudication, depending upon whether they perceived they would be better or worse off as a result. It is thus, inevitably, a compromise solution.

Bearing that in mind, although the Scheme has been adversely criticised by many, it is invariably better than the small number of other formal adjudication procedures that existed before the Construction Act. There is no doubt that the Scheme formed a useful checklist and benchmark against which later Act-based contractual procedures were developed. In fact, it is the preferred option for a small number of users over other published alternatives.

Following substantial consultation, 'the Scheme' relating to England is currently thought likely to be amended by legislation in the relatively near future. (The Schemes relating to Wales, Scotland and Northern Ireland may have their own amendments.) For the purposes of this text, areas of the Scheme applicable to England which may be amended are referred to at the appropriate places within this chapter for ease of reference, but are discussed in more detail in Chapter 8.

The Scheme can be used in two principal ways. Firstly, as intended, it is available as an automatic fall-back option for a Referring Party if a contract's existing adjudication provisions do not comply with the Construction Act. Secondly, it can be selected as the preferred adjudication system anyway (i.e. it can be expressly referred to, or its wording included, as the pertaining adjudication procedures within conditions of contract), in the same way as any other adjudication procedures. In the latter circumstances, there is also no restriction on making amendments to it to more readily reflect the requirements of the drafting Party or construction project, to sharpen up perceived weaknesses and so on. However, if the amendments are written into the contract documents and do not comply with what is written in the Construction Act, a Referring Party in a contract coming under the Construction Act would once again would presumably have the option to select the unaltered Scheme to resolve its dispute!

The Adjudication provisions

Paragraph 1 (1) Notice of intention to refer to adjudication

Any party to a construction contract (the 'referring party') may give written notice (the 'notice of adjudication') of his intention to refer any dispute arising under the contract, to adjudication.

Paragraph 1 (2) Notice to be given to all Parties

The notice of adjudication shall be given to every other party to the contract.

It will be noted that 'other' Parties to the contract (i.e. not just those to the dispute) might include guarantors and providers of bonds (even though these are excluded from being construction contracts under the Construction Act by the Secretary of State) if they are signatories to the contract concerned. Insurers and financiers might also be involved. The legal consequences of not complying with this are not known.

Paragraph 1 (3) Details to be set out

The notice of adjudication shall set out briefly –

(a) the nature and a brief description of the dispute and of the parties involved,

(b) details of where and when the dispute has arisen,

(c) the nature of the redress which is sought, and

(d) the names and addresses of the parties to the contract (including, where appropriate, the addresses which the parties have specified for the giving of notices).

The Notice of Adjudication must include brief details of the dispute, the Parties (with addresses for communications) and the 'nature of redress sought'. In fact, the definition of a dispute and the extent to which it is set out in the Notice of Adjudication can be critical. This is considered in more detail in Chapter 14 Miscellaneous Issues.

Paragraph 2 (1) Request for appointment of Adjudicator

Following the giving of a notice of adjudication and subject to any agreement between the parties to the dispute as to who shall act as Adjudicator –

(a) the referring party shall request the person (if any) specified in the contract to act as Adjudicator, or

(b) if no person is named in the contract or the person named has already indicated that he is unwilling or unable to act, and the contract provides for a specified nominating body to select a person, the referring party shall request the nominating body named in the contract to select a person to act as Adjudicator, or

(c) where neither paragraph (a) nor (b) above applies, or where the person referred to in (a) has already indicated that he is unwilling or unable to act and (b) does not apply, the referring party shall request an Adjudicator nominating body to select a person to act as Adjudicator.

Following the issue of the Notice of Adjudication, the Referring Party must request any pre-agreed or specified named Adjudicator, or specified nominating body, or non-specified Adjudicator nominating body, in that order, to be or to select an Adjudicator for the particular dispute. In other words, the Paragraph provides a 'cascading' provision for the appointment of an Adjudicator so as to ensure the appointment in all reasonably conceivable circumstances.

It will be noted that this is one area where the wishes of the Parties, either before or after the dispute, or by provision within the conditions of contract, are expressly permitted to take precedence before the statutory provisions of the Scheme start applying.

Paragraph 2 (2) Confirmation by Adjudicator

A person requested to act as Adjudicator in accordance with the provisions of paragraph 2 (1) shall indicate whether or not he is willing to act within two days of receiving the request.

It is at this stage that a potential Adjudicator will need to consider whether there are issues such as any conflicts of interest, whether it will be possible to act impartially and whether diary commitments allow for a decision to be produced within 28 days.

It is also at this point that the Adjudicator, if it has not already done so, will need to send to the Parties its terms and conditions. There should be little, or no trouble in gaining the support of the courts if terms and conditions are subject to their own dispute later, as long as the latter are:

(a) Reasonable.
(b) Do not conflict with the Construction Act's aim for a rapid economic resolution of a dispute.
(c) Not expressly mentioned in the Construction Act or Scheme.
(d) Expressly mentioned in the Construction Act or Scheme but are not at variance with those requirements.
(e) At variance with the express requirements of the Construction Act or Scheme, but the Parties both clearly agree to waive those express requirements for this particular dispute.

It will be noted that a proposal or a requirement, for example, for an Advance Payment to be made as security, or part security, for the Adjudicator's fees and expenses would come into several of the foregoing categories, e.g. many Adjudicators would believe that some provision for the security of their fees and expenses was important and perfectly reasonable. This does not prevent the Parties refusing to agree to a proposal to pay an Advance, but that is merely between them and the Adjudicator, not one related to the Construction Act or Scheme. In this case, disagreement on this issue can, at its worst, mean that the Adjudicator refuses any further involvement with the dispute, otherwise the Adjudicator must continue without an agreement. Terms and conditions are discussed in more detail in Chapter 9 Adjudicator Appointments.

For the Parties, this Paragraph enables them to know within 2 days whether they have secured the appointment of an Adjudicator or whether they have to continue with the next stage of the appointment process. If the latter, and the Section 108 (2)(b) of the Construction Act timetable requirement to both appoint an Adjudicator and refer the dispute to that Adjudicator within 7 days of the Notice of Adjudication is considered mandatory (although, as can be seen from the commentary on Section 108 (2)(b), it is not thought to be so), the Referring Party might consider issuing (or the need to issue) a new Notice of Adjudication in order to comply with that requirement.

Paragraph 2 (3) Adjudicator Nominating Body

In this paragraph, and in paragraphs 5 and 6 below, an 'Adjudicator nominating body' shall mean a body (not being a natural person and not being a party to the dispute) which holds itself out publicly as a body which will select an Adjudicator when requested to do so by a referring party.

i.e. an Adjudicator Nominating Body (ANB) may be any corporate body, other than one of the Parties, which holds itself out publicly to appoint Adjudicators. Interestingly there are no statutory or other regulations or qualifications governing ANBs, so any organisation can be one.

It has also been suggested that the concept is not totally satisfactory as the choice of ANB is left unilaterally to the Referring Party. However, as an Adjudicator must act impartially, in theory that should not be a problem.

Further, as any corporate body can be an ANB and the government has deliberately chosen not to regulate ANBs or Adjudicators (e.g. the government plays no part in their training or qualification standards), there can be a question mark over the quality of Adjudicator proposed. Again however, an ANB is likely to be associated with a specialised disputes organisation, or disputes division of an organisation, the members of which will probably have obtained their adjudication training from recognised national training courses, or from those who have been on such courses. In practice, a non-specified ANB might not need to be approached anyway, as it is a final resort following approaches to pre-agreed or specified Adjudicators, or to a specified nominating body.

Paragraph 3 Copy of Notice of Adjudication to be supplied

> *The request referred to in paragraphs 2, 5 and 6 shall be accompanied by a copy of the notice of adjudication.*

i.e. a copy of the Notice of Adjudication must be supplied with the request to act as, or request to select, an Adjudicator in accordance with Paragraph 2 (1). This will permit a potential Adjudicator to assess whether the dispute is of the type it can decide.

Paragraph 4 Adjudicator to be 'natural person', not employee, and declare interest

> *Any person requested or selected to act as Adjudicator in accordance with paragraphs 2, 5 or 6 shall be a natural person acting in his personal capacity. A person requested or selected to act as an Adjudicator shall not be an employee of any of the parties to the dispute and shall declare any interest, financial or otherwise, in any matter relating to the dispute.*

An Adjudicator cannot therefore be an organisation (e.g. a company, or a university department). This implies that an Adjudicator's agreement should be between the Adjudicator itself and the Parties. If this is correct, where the Adjudicator is someone's employee, protected by the Professional Indemnity (PI) insurance cover and provided with other support by their parent organisation, separate arrangements outside the agreement, in theory at least, will need to be made by the Adjudicator with that organisation regarding issues such as the reimbursement of the Adjudicator's fees and expenses. The foregoing requirement seems unnecessarily restrictive. In practice, Adjudicators who are employees are usually making that fact clear to Parties and are requesting that fee payments are made to their employer's accounts. Indeed, the courts (*Faithful & Gould* v *Arcal* [2001]) agree that an Adjudicator does not need to sue for its fees as a natural person acting in a personal capacity. These types of practical procedures should clearly make no difference to the adjudication process or the quality or timing of the resulting adjudication decision.

Paragraph 5 (1) Proposal within 5 days

The nominating body referred to in paragraphs 2 (1)(b) and 6 (1)(b) or the Adjudicator nominating body referred to in paragraphs 2 (1)(c), 5 (2)(b) and 6 (1)(c) must communicate the selection of an Adjudicator to the referring party within five days of receiving a request to do so.

i.e. a specified nominating body or ANB must notify the Referring Party of the selected Adjudicator within 5 days of the request. This allows the Parties to know where they are relative to the overall 7-day timetable requirement of Section 108 (2)(b) of the Construction Act.

Paragraph 5 (2) Failure to appoint

Where the nominating body or the Adjudicator nominating body fails to comply with paragraph (1), the referring party may –

(a) agree with the other party to the dispute to request a specified person to act as Adjudicator, or

(b) request any other Adjudicator nominating body to select a person to act as Adjudicator.

i.e. if a nominating body or ANB does not select an Adjudicator within the 5 days allowed, then the Referring Party has to start again, this time being limited to either agreeing a proposed nominee with the other Party or by going to another (note: not the same) ANB.

It will be apparent that the failure to select an Adjudicator will put the Section 108 (2)(b) of the Construction Act timetable requirement to both appoint an Adjudicator and refer the dispute to that Adjudicator within 7 days of the Notice of Adjudication significantly out of time. If this is considered to be mandatory (although, as can be seen from the commentary on Section 108 (2), it is not thought to be so), the Referring Party will have no choice if it wants to continue with adjudication but to issue a new Notice of Adjudication in order to attempt again to comply with that requirement.

Paragraph 5 (3) Adjudicator to communicate

The person requested to act as Adjudicator in accordance with the provisions of paragraphs 5 (1) or 5 (2) shall indicate whether or not he is willing to act within two days of receiving the request.

This is almost identical to Paragraph 2 (2).

Paragraphs 6 (1) and 6 (2) Named Adjudicator

(1) Where an Adjudicator who is named in the contract [as described in paragraph 2 (1)(a)] *indicates to the parties that he is unable or unwilling to act, or where he fails to respond in accordance with paragraph 2 (2), the referring party may –*

> *(a) request another person (if any) specified in the contract to act as Adjudicator, or*
>
> *(b) request the nominating body (if any) referred to in the contract to select a person to act as Adjudicator, or*
>
> *(c) request any other Adjudicator nominating body to select a person to act as Adjudicator*
>
> *(2) The person requested to act in accordance with the provisions of paragraph [6](1) shall indicate whether or not he is willing to act within two days of receiving the request.*

i.e. any named potential Adjudicator as described in Paragraph 2 (1)(a) has 2 days to indicate a willingness to act. If not (or in the absence of a response in 2 days), the Referring Party must request any other specified person to act as Adjudicator, or request any specified nominating body or another ANB to select a person.

This is very similar in terms of words to that given earlier in Paragraph 2 (1). Should the potential Adjudicator not respond within 2 days, the comments given under that Paragraph relating to the Construction Act's timetable provisions will again apply.

Paragraph 7 (1) Referral notice to Adjudicator

> *Where an Adjudicator has been selected in accordance with paragraphs 2, 5 or 6, the referring party shall, not later than seven days from the date of the notice of adjudication, refer the dispute in writing (the 'referral notice') to the Adjudicator.*

This completes the Section 108 (2)(b) of the Construction Act timetable for the referral of the dispute to the Adjudicator within 7 days of the Notice of Adjudication.

As discussed under Section 108 (2)(b) in Chapter 2 and under the heading 'Urgency of appointment' in Chapter 9, it is not clear whether Adjudicator appointment and referral within 7 days is a mandatory requirement of the Construction Act or not. Paragraph 7 (1) makes it an express requirement (and thus appears to see no need to state any other procedures to be followed if the 7 day timetable is not achieved, other than by implication, to start the process again). Until and unless the courts clarify the intention of the Construction Act, this seems an unnecessarily onerous express requirement that will sometimes not be possible to meet.

Paragraph 7 (2) Documents with referral notice

> *A referral notice shall be accompanied by copies of, or relevant extracts from, the construction contract and such other documents as the referring party intends to rely upon.*

These should be comprehensive because there may be no opportunity to add to them later.

Paragraph 7 (3) Other Parties to receive notice and documents

> *The referring party shall, at the same time as he sends to the Adjudicator the documents referred to in paragraphs [7](1) and [7](2), send copies of those documents to every other party to the dispute.*

The other Parties must be sent copies of the referral notice and documents at the same time as the Adjudicator. The legal consequences of not doing so are not known.

It is noted that there is nothing in Paragraph 7 or elsewhere that refers to any response by the Responding Party or a time by which it should be received by the Adjudicator and Referring Party. (Equally, there is nothing in the Scheme that prevents a Responding Party from providing a response.)

Words similar to those used in the ICE Procedure could have been used, e.g.:

> *Unless otherwise agreed between the Adjudicator and both Parties, the Responding Party shall, if it has a response to the Referring Party's case, submit that response within 14 days of the Date of Referral, or earlier if instructed by the Adjudicator.*

In some circumstances, the Parties might consider agreeing to supplement the Scheme wording to incorporate issues such as this.

Paragraph 8 (1) More than one dispute at same time

> *The Adjudicator may, with the consent of all the parties to those disputes, adjudicate at the same time on more than one dispute under the same contract.*

This provision would include disputes in the same contract on related issues. The Paragraph means that an Adjudicator cannot adjudicate more than one dispute at a time arising under a single contract without the consent of the other Party. Thus, counter-claims not already raised as such as a defence in an adjudication but raised as a separate adjudication, would have to be decided by a different Adjudicator (if one of the Parties did not agree to the same Adjudicator). Put another way, if one Party objects, a counter-claim cannot be referred to its own adjudication until a decision has been reached on the first one, if the same Adjudicator was required.

Paragraph 8 (2) Related disputes

> *The Adjudicator may, with the consent of all the parties to those disputes, adjudicate at the same time on related disputes under different contracts, whether or not one or more of those parties is a party to those disputes.*

i.e. an Adjudicator can (if it wants to) adjudicate on related disputes under different contracts at same time.

This permits an Adjudicator to adjudicate on related main contract and sub-contract disputes, but only if the Parties to the relevant contract and sub-contracts agree. This appears to be notwithstanding whether those other disputes were initially properly referred to the Adjudicator by the adjudication procedure nomination process. Joining of disputes is discussed in more detail in Chapter 14, Miscellaneous issues.

Paragraph 8 (3) Agreement to extend time

> *All the parties in paragraphs* [8] *(1) and* [8] *(2) respectively may agree to extend the period within which the Adjudicator may reach a decision in relation to all or any of these disputes.*

It will be noted that the Adjudicator is not apparently involved in this Paragraph. However, in practice, the Adjudicator will probably only usually agree to become involved in any Paragraph 8 (1) or 8 (2) additional disputes if the Paragraph 8 (3) issue is resolved to its satisfaction at the same time.

Paragraph 8 (4) Adjudicator ceases to act, fees

> *Where an Adjudicator ceases to act because a dispute is to be adjudicated on by another person in terms of this paragraph, that* [former] *Adjudicator's fees and expenses shall be determined in accordance with paragraph 25.*

Where the Parties agree to combine existing disputes and adjudications under one Adjudicator then the one(s) that have to step down are entitled to be reimbursed fees and expenses incurred to that point.

Paragraph 9 (1) Resignation at any time

> *An Adjudicator may resign at any time on giving notice in writing to the parties to the dispute.*

The value of this express provision, is not usually found in other adjudication procedures, is not clear i.e. if this Paragraph was omitted, would it make any difference in practise?.

Paragraph 9 (2) Obligatory resignation

> *An Adjudicator must resign where* [it becomes apparent that] *the dispute is the same or substantially the same as one which has previously been referred to adjudication, and a decision has been taken in that adjudication.*

This Paragraph assists provide temporary finality to an adjudication decision subject only to final determination in accordance with Section 108 (3) of the Construction Act. It appears on its face to apply even where a previous adjudication decision has been successfully overturned by the courts after an appeal, as the trigger for resignation is a decision on the previous dispute, something that is not affected by the latter's merits. This seems to be overly onerous and unnecessary. An alternative approach is to take such a court decision as destroying the previous adjudication decision, so that there would be no previous adjudication decision for a subsequent Adjudicator to resign over.

Paragraph 9 (3) Action on resignation

Where an Adjudicator ceases to act under paragraph 9 (1) –

(a) the referring party may serve a fresh notice under paragraph 1 and shall request an Adjudicator to act in accordance with paragraphs 2 to 7; and

(b) if requested by the new Adjudicator and insofar as it is reasonably practicable, the parties shall supply him with copies of all documents which they had made available to the previous Adjudicator.

When an Adjudicator resigns in accordance with Paragraph 9 (1), the Referring Party may, if it wants to, serve a 'fresh' written Notice of Adjudication and start the Adjudicator selection process all over again. As far as reasonably possible, if the new Adjudicator asks, the Parties must then provide all the documents supplied to the first Adjudicator. The dispute must, of course, remain the one originally referred (unless the Parties and Adjudicator can agree to include other matters under Paragraph 20 (see later)).

However, having a new Adjudicator presents a Referring Party with an opportunity to take stock and possibly revise its case, particularly in the light of seeing the position being taken by, and the documents of, the Responding Party. This might result in an entirely new, different, referral being initiated before the new Adjudicator selection processes are commenced.

Paragraph 9 (4) Action on resignation

Where an Adjudicator resigns in the circumstances referred to in paragraph [9] (2), or where a dispute varies significantly from the dispute referred to him in the referral notice and for that reason he is not competent to decide it, the Adjudicator shall be entitled to the payment of such reasonable amount as he may determine by way of fees and expenses reasonably incurred by him. The parties shall be jointly and severally liable for any sum which remains outstanding following the making of any determination on how the payment shall be apportioned.

This preserves the fees and expenses of an Adjudicator that has unfortunately found itself in the situation described in Paragraph 9 (1).

It will be noted that who decides on the Adjudicator's competency is not stated e.g. it is one or both Parties, or is it an Adjudicator's opinion only.

Paragraph 10 Objection to appointment to have no effect

Where any party to the dispute objects to the appointment of a particular person as Adjudicator, that objection shall not invalidate the Adjudicator's appointment nor any decision he may reach in accordance with paragraph 20.

The statement, on its face, is very clear. It would apply, for instance, where a Party objects to an Adjudicator appointed under Paragraph 2 (1), because it considered that Adjudicator to be inappropriately experienced or more likely to be sympathetic to the views of the other Party. It will be noted that there is no restriction on when the objection can be made, i.e. at the time of appointment or later. Interestingly, it would also apply if an Adjudicator had not stated any possible conflict of interest in accordance with Paragraph 4 at the time of its appointment, as even if such a conflict had been declared, the most a Party could have done was object, and that is covered by this Paragraph.

However, a Party might object to an Adjudicator's continuing appointment if that Party believes that the Adjudicator was, for example, acting partially, contrary to Paragraph 12, or that the Adjudicator was considering, or finally deciding on, matters not referred to it. Although under this Paragraph an objection itself cannot invalidate an appointment or decision, this does not prevent the grounds for the objection being used to attempt to later invalidate that appointment or decision in court. Indeed, any such objection made in court would be strengthened by it also having been made earlier in the adjudication.

Paragraph 11 (1) Revocation of appointment by Parties

The parties to a dispute may at any time agree to revoke the appointment of the Adjudicator. The Adjudicator shall be entitled to the payment of such reasonable amount as he may determine by way of fees and expenses incurred by him. The parties shall be jointly and severally liable for any sum which remains outstanding following the making of any determination on how the payment shall be apportioned.

The Scheme is very detailed in Paragraphs 8 (4), 9 (4) and 11 (1) concerning payment of Adjudicator fees and expenses in certain identified situations when an Adjudicator's role prematurely terminates. In fact, the Scheme is more comprehensive on this issue than most other adjudication procedures, which do not expressly permit the Parties to bring an adjudication to an end before an adjudication decision has been reached.

Paragraph 11 (2) Revocation because of default or misconduct

Where the revocation of the appointment of the Adjudicator is due to the default or misconduct of the Adjudicator, the parties shall not be liable to pay the Adjudicator's fees and expenses.

Misconduct is dealt with in more detail in Chapter 11, Procedural fairness and in Chapter 14, Miscellaneous issues.

Paragraph 12 Adjudicator to act impartially and avoid unnecessary expense

The Adjudicator shall –

(a) act impartially in carrying out his duties and shall do so in accordance with any relevant terms of the contract and shall reach his decision in accordance with the applicable law in relation to the contract; and

(b) avoid incurring unnecessary expense.

An Adjudicator must act impartially, avoid unnecessary expense and reach a decision in accordance with the applicable law in relation to the contract. It will be noted that there is no requirement for the Adjudicator to be independent as required by the European Convention on Human Rights or some contractual adjudication procedures. This is considered in more detail in Chapter 11 Procedural fairness.

Paragraph 13 Adjudicator may take initiative

The Adjudicator may take the initiative in ascertaining the facts and the law necessary to determine the dispute, and shall decide on the procedure to be followed in the adjudication. In particular he may –

(a) request any party to the contract to supply him with such documents as he may reasonably require including, if he so directs, any written statement from any party to the contract supporting or supplementing the referral notice and any other documents given under paragraph 7 (2),

(b) decide the language or languages to be used in the adjudication and whether a translation of any document is to be provided and if so by whom,

(c) meet and question any of the parties to the contract and their representatives,

(d) subject to obtaining any necessary consent from a third party or parties, make such site visits and inspections as he considers appropriate, whether accompanied by the parties or not,

(e) subject to obtaining any necessary consent from a third party or parties, carry out any tests or experiments,

(f) obtain and consider such representations and submissions as he requires, and, provided he has notified the parties of his intention, appoint experts, assessors or legal advisers,

(g) give directions as to the timetable for the adjudication, any deadlines, or limits as to the length of written documents or oral representations to be complied with, and

(h) issue other directions relating to the conduct of the adjudication.

An Adjudicator can restrict the length of written documents to be submitted by the Parties. The most obvious advantage to an Adjudicator would be to limit the size/extent of the bundle of referral documents. That would also concentrate the Referring Party's mind and reduce the extent of possible 'ambushing' (i.e. the provision of large amounts of documents that might be impractical to consider properly in the available time). Unfortunately, an Adjudicator has no control over an adjudication until after appointment, i.e. too late in practice to instruct a Referring Party to limit its size of its initial submission. If the latter has forwarded a large number of documents in support of its case, it would be partial to impose any limitation on any later bundle from a Responding Party. One possible way around this is for the Adjudicator to ask the Referring Party to resubmit its case in a more limited precise form, with the support provided by Paragraphs 14 and 15 (following). This issue is discussed in more detail in Chapter 12 Conduct of the Adjudication.

Paragraphs 14 and 15 Requests by Adjudicator

The parties shall comply with any request or direction of the Adjudicator in relation to the adjudication.

If, without showing sufficient cause, a party fails to comply with any request, direction or timetable of the Adjudicator made in accordance with his powers, fails to produce any document or written statement requested by the Adjudicator, or in any other way fails to comply with a requirement under these provisions relating to the adjudication, the Adjudicator may –

(a) continue the adjudication in the absence of that party or of the document or written statement requested,

(b) draw such inferences from that failure to comply as circumstances may, in the Adjudicator's opinion, be justified, and

(c) make a decision on the basis of the information before him attaching such weight as he thinks fit to any evidence submitted to him outside any period he may have requested or directed.

Non-compliance may, in practice, result in a more adverse decision where a Party is inefficient or deliberately decides it is more in its interest not to comply with an Adjudicator's request or instruction. These Paragraphs are thus a very useful weapon in an Adjudicator's armoury. An obvious potential situation is where one Party refuses to have anything to do with an adjudication and will not communicate further with an Adjudicator. In addition, they can be used, for example, to curtail potential injustices that might arise from a Referring Party attempting to 'ambush' a Responding Party with a large amount of later referral documentation.

Paragraph 16 (1) Legal and other assistance to Parties

Subject to any agreement between the parties to the contrary, and to the terms of paragraph [16]*(2) below, any party to the dispute may be assisted by, or represented by, such advisers or representatives (whether legally qualified or not) as he considers appropriate.*

Each Party is free to be represented by others, without obtaining the agreement of the other Party or the Adjudicator.

Paragraph 16 (2) Party representation at Adjudicator meetings

Where the Adjudicator is considering oral evidence or representations, a party to the dispute may not be represented by more than one person, unless the Adjudicator gives directions to the contrary.

This, in practice, would cover all evidential and fact finding meetings, but exclude any meetings that an Adjudicator feels are necessary (and has the time to attend) where it is solely giving instructions, discussing its appointment, etc. One result should be to reduce the costs of the adjudication.

Paragraph 17 Relevant information to Parties

The Adjudicator shall consider any relevant information submitted to him by any of the parties to the dispute and shall make available to them any information to be taken into account in reaching his decision.

The latter refers to information relevant to the decision which has not been supplied to a Party during the adjudication by the Adjudicator or the other Party. It would include the Adjudicator's own knowledge and experience where that is relevant to the Adjudicator's decision.

The apparent obligation (although this is not interpreted as such by some commentators) to consider all relevant information whenever supplied before a decision is reached is seen to be a potential open door to 'ambushing'. The latter is discussed in more detail under the headings of 'Limiting the extent of the referral documents', and 'Ambushing' in Chapter 12.

Essentially, solutions consist of the Adjudicator requiring extensions of time and/or summaries in order for it and the Responding Party to have reasonable time to deal with the documents. The former, at least, requires a Party or the Parties to agree and such agreement might not be readily forthcoming. In fact, a robust Party might consider refusing just to have grounds for an appeal on the basis of an infringement of the obligatory nature of Paragraph 17 (i.e. the Adjudicator did not/could not comply with the 'shall consider any' (equals all?) obligation of the Paragraph), should that Party not be satisfied with the decision. (See also Chapter 11, Procedural fairness.) It will be noted also that Paragraph 17 taken with Paragraph 12 might well be deemed to be equivalent to an express natural justice requirement, even if the words 'natural justice' are not used.

Some commentators have suggested deleting Paragraph 17 from the Scheme altogether to overcome such potential problems. Another alternative would be some form of clearer rewording, perhaps incorporating extensions of time and/or requiring summaries to be provided.

Paragraph 18 Confidentiality

The Adjudicator and any party to the dispute shall not disclose to any other person any information or document provided to him in connection with the adjudication which the party supplying it has indicated is to be treated as confidential, except to the extent that it is necessary for the purposes of, or in connection with, the adjudication.

This provision enables a Party, concerned about information submitted in an adjudication, to know that it will be treated confidentially. It will not necessarily prevent that information being used in other adjudications, subsequent litigation or arbitration between the same Parties. For example, if enforcement of a decision or a final determination goes to court, documents such as a decision will become a matter of court record and therefore be public.

Paragraph 19 (1) Time for decision

The Adjudicator shall reach his decision not later than –

(a) twenty eight days after the date of the referral notice mentioned in paragraph 7 (1), or

(b) forty two days after the date of the referral notice if the referring party so consents, or

(c) such period exceeding twenty eight days after the referral notice as the parties to the dispute may, after the giving of that notice, agree.

It would have been better if, instead of the phrase 'date of the referral notice', the phrase 'actual date of receipt of referral documents by the Adjudicator' had been used.

It will be noted that a decision has to be reached, not necessarily supplied to the Parties, by the foregoing time limits. However, Paragraph 19 (3) (following) requires the decision to be provided to the Parties 'as soon as possible' after being reached, so this eliminates any scope for, say, the Adjudicator retaining its decision until it receives payment of its fees and expenses.

Paragraph 19 (2) Failure to meet time requirement

Where the Adjudicator fails, for any reason, to reach his decision in accordance with paragraph [19] *(1) –*

(a) any of the parties to the dispute may serve a fresh notice under paragraph 1 and shall request an Adjudicator to act in accordance with paragraphs 2 to 7; and

(b) if requested by the new Adjudicator and insofar as it is reasonably practicable, the parties shall supply him with copies of all documents which they had made available to the previous Adjudicator.

It is noted that the foregoing is an option only. The Parties are not precluded from accepting a late Adjudicator decision (or, presumably, not accepting it, but deciding also not to serve a new notice etc.). A potential problem might be that one Party accepts a late decision, but the other refuses to accept it; in this case the decision would probably be a nullity. It is considered unlikely that an Adjudicator of an unaccepted late decision would be paid its fees.

Paragraph 19 (3) Decision to be provided to Parties

As soon as possible after he has reached a decision, the Adjudicator shall deliver a copy of that decision to each of the parties to the contract.

This precludes an Adjudicator retaining a decision until payment of its fees and expenses.

This does not prevent an Adjudicator from trying to include in its initially proposed terms and conditions, or proposing later during the adjudication, for such payment of its fees and expenses. It means that if the Parties disagree to the proposal, the Adjudicator has no authority to withhold a decision once it has been made. It is suggested that the implications of such a proposal should be made clear to 'unsophisticated' Parties (certainly those not legally represented) to avoid the possibility of complaints that might be made, not least to the Adjudicator's appointing body, if a Party considers it has been misled.

Paragraph 20 Details of decision

The Adjudicator shall decide the matters in dispute. He may take into account any other matters which the parties to the dispute agree should be within the scope of the adjudication or which are matters under the contract which he considers are necessarily connected with the dispute.

[Note that current indications are that future legislative changes to the Scheme may be introduced relating to this issue, stating the following, or similar:

but shall not take into account any matter relating to the legal or other costs of the parties arising out of or in connection with the adjudication.

This precludes an Adjudicator deciding on Party costs. Possible legislative Scheme Amendments are discussed in more detail in Chapter 8.]

In particular, he may –

(a) open up, revise and review any decision taken or any certificate given by any person referred to in the contract unless the contract states that the decision or certificate is final and conclusive[.]

The initial preamble permits the Parties and Adjudicator to agree to consider matters that were not included as part of an initial referral. Issues such as time remaining and Adjudicator availability will affect that decision. The Adjudicator is permitted to unilaterally decide whether to include any such connected matters.

Sub-paragraph (a) states that an Adjudicator is not allowed to open up, review and revise any decision or certificates said to be final and conclusive in the contract. (This should, perhaps, be distinguished from final and conclusive agreements or settlements reached outside the contract. The latter are discussed under the heading 'Can an Adjudicator review all certificates and decisions' in Chapter 14.) The sub-paragraph gives a contract drafting Party not wishing to use (or let another Party use) adjudication to change, say, any certificate (i.e. even those that are 'interim' ones, not just the last or 'final' certificate), a way to avoid adjudication on a dispute on such certificates, just by stating in a contract that 'interim' certificates are to be taken as final and conclusive. It is not clear why in practice a certifying Employer might wish to do this, and care would be needed in wording to ensure that this did not preclude reassessment and adjustment, if completed certified works were later found to be defective.

The sub-paragraph caused substantial comment when it was first published, as it seemed an easy back-door way for disputes on *all* previous appropriately worded decisions or certificates to avoid adjudication under the Construction Act and that appeared to be directly contrary to what that Act was aiming at. In fact, the sub-paragraph is being realistic. The initial criticism ignored the fact that 'final and conclusive' agreements, which could not be altered even by the courts, occurred in some circumstances in practice anyway and this needed to be acknowledged by the Scheme. Points in favour of the sub-paragraph include:

- Principal purposes of the Construction Act include to reduce the impact of disputes and to reach their quick and economic resolution. Therefore, any contract that had conclusive clauses at any stage, e.g. whether within the general clauses (say, resulting in a fixed price contract, or one where risks were clearly defined, such as omitting unexpected ground condition clauses) or within any contractual ADR provisions (say, expert determination, Engineer's decision or mediation) would presumably be very much in the spirit of the Act. If this finality occurred before an adjudication stage was reached, then surely so much the better.
- The courts themselves (e.g. the House of Lords in *Beaufort* v *Ash* [1999] and the associated overruling of the Court of Appeal decision in *Northern Regional Health Authority* v *Crouch* [1984]) cannot open up, review and revise any decision expressly stated to be final and conclusive in the contract. Why therefore should Adjudicators have that right?
- Certain ADR provisions, such as expert determination, may be clearly and expressly stated in the construction contract to be final and conclusive, as agreed and intended between the Parties. There would seem no question that Adjudicators can overrule such agreements.
- Adjudication decisions themselves can be agreed between the Parties to be final and binding (see Chapter 2 on the Construction Act, Section 108 (3)). Why should final and conclusive adjudication decisions be subject to further adjudication when other final and conclusive ADR

provisions are not? In fact, as already noted in Paragraph 9 (2), an Adjudicator is not allowed to rehear a previously decided dispute.

Paragraph 20 therefore acknowledges the advantages and existence already of express 'final and conclusive' agreements and no alternative adjudication procedures have been seen which better address this issue.

Paragraph 20 continues, giving an Adjudicator powers to:

(b) decide that any of the parties to the dispute is liable to make a payment under the contract (whether in sterling or some other currency) and, subject to section 111 (4) of the [Construction] *Act, when that payment is due and the final date for payment,*

(c) having regard to any term of the contract relating to the payment of interest decide the circumstances in which, and the rates at which, and the periods for which simple or compound rates of interest shall be paid.

On its face this seems unambiguous. However, commentators interpret (c) either as 'Any contract provision must be followed, thus if there is no provision for interest in the contract, then the Adjudicator cannot award it', or alternatively that 'the Adjudicator can award interest notwithstanding what is stated in the contract'. The former seems more logical, otherwise why include the phrase 'having regard to any term in the contract relating to the payment of interest ...'. If not clarified in a future amendment, this matter will probably need to be decided by the courts (see also 'Interest' in Chapter 13).

Paragraph 21 Immediate compliance

In the absence of any directions by the Adjudicator relating to the time for performance of his decision, the parties shall be required to comply with any decision of the Adjudicator immediately on delivery of the decision to the parties in accordance with this paragraph.

In other words, the Parties shall comply immediately with any received decision, unless otherwise directed by the Adjudicator.

Paragraph 22 Reasons

If requested by one of the parties to the dispute, the Adjudicator shall provide reasons for his decision.

It will be noted that it is not stated when a request for reasons should be made. In fact, this should be substantially in advance of the date for reaching a decision. That then enables the Adjudicator to better plan its input in the limited time still available to it so that the reasons can be provided with the decision. It is suggested that the Parties are informed that reasons will be provided to both of them, even though only one Party may have requested them.

Published reasons are generally not favoured by Adjudicators because firstly, they can take up limited available time and secondly, they can provide grounds for disputing the decision and result in a possible challenge. Lawyers generally would obviously like reasons for that very purpose. Insurers also want them to assist in distinguishing insured from uninsured elements of a decision and, additionally, to be better able to assess whether the funding of a later arbitration or litigation can be justified.

Parties who want finality of decision would be against reasons, but those that consider adjudication an unavoidable fact finding and strength of case testing aggravation prior to arbitration or litigation, or simply want to know details so that they can, for example, correct organisational shortcomings, would not.

Current indications are that future legislative changes to the draft Scheme may replace Paragraph 22 with the following or similar:

> *22 (1) Subject to paragraph* [22] *(2), if requested by one of the parties, the Adjudicator shall provide reasons for his decision.*
>
> *22 (2) The Adjudicator may set a deadline for the parties to request the reasons for his decision and where a request is submitted outside that deadline, the Adjudicator shall not be required to comply with paragraph* [22] *(1).*

This permits an Adjudicator to state when a request for reasons must be given. Draft Scheme Amendments are discussed in more detail in Chapter 8.

Reasons are discussed in more detail in Chapter 13.

New Paragraph 22A Corrections

Possible future legislative changes to the Scheme to deal with Adjudicator corrections are as follows:

> *22A (1) The Adjudicator may, on his own initiative or on the application of a party, correct his decision so as to remove any clerical mistake or error arising from an accidental slip or omission.*
>
> *22A (2) Any application for the exercise of the Adjudicator's powers under paragraph* [22A] *(1) shall be made within 5 days of the date that the decision is delivered to the parties or such shorter period as the Adjudicator may specify in his decision.*
>
> *22A (3) Any correction of a decision shall be made as soon as possible after the date that the application was received by the Adjudicator or, where the correction is made by the Adjudicator on his own initiative, as soon as possible after he becomes aware of the need to make a correction.*

This allows an Adjudicator to correct some errors in its decision. Possible legislative Scheme Amendments are discussed in more detail in Chapter 8.

Paragraph 23 (1) Option for peremptory compliance

In his decision, the Adjudicator may, if he thinks fit, order any of the parties to comply peremptorily with his decision or any part of it.

An Adjudicator may or may not order peremptory compliance. Peremptory compliance enables the courts to issue an order to comply, failure to do so giving rise to the consequences for a breaching a court order, such as fines or imprisonment for contempt of court. In fact, a peremptory order may normally be too heavy-handed, the fact of an adjudication decision being made (possibly reinforced in some cases by express requirements for immediate compliance, such as that in Paragraph 21) constituting grounds enough for an early court enforcement action. The usual procedure following an adjudication decision, particularly in money cases, would be for a Party to ignore any peremptory order procedure even if the option exists, and just issue court proceedings claiming the amount due. This would be followed by an application to the courts for summary judgment, all in accordance with normal debt recovery procedures. This procedure could also be enhanced by a Party requesting a court to reduce the timetable for normal debt recovery.

For non-monetary cases, it has been suggested that the correct course of action is to apply for a mandatory injunction. It would therefore appear that this Paragraph would benefit from revision to reflect what is actually happening in the courts to enforce adjudication decisions.

Peremptory compliance and enforcement are also similarly discussed in Chapter 13 and Chapter 15 respectively.

Paragraph 23 (2) Binding decision

The decision of the Adjudicator shall be binding on the parties, and they shall comply with it until the dispute is finally determined by legal proceedings, by arbitration (if the contract provides for arbitration or the parties otherwise agree to arbitration) or by agreement between the parties.

The courts have robustly supported this view, although they will look carefully at challenges based on allegations that an Adjudicator has not got, or has exceeded, its jurisdiction. This issue is discussed in more detail in Chapter 10.

Paragraph 23 (2) is thus consistent with the requirement of Paragraph 9 (2) that an Adjudicator cannot rehear what is essentially the same dispute. If that was not the case, either the previous decision would not be binding as required by this Paragraph, or there might be two different but binding decisions on the same issue. This unfortunately appears to prevent a new adjudication on an ongoing developing issue. For example, a Referring Party might require interim monetary decisions relating to a problem occurring on site over a period of months, the true costs of which might only be known when that problem was finally resolved. In some circumstances, it might be possible to agree between the Parties that further adjudication was part of the permitted 'agreement between the Parties' phrase. Otherwise, it would appear that 'agreement between the Parties' could be reached with the assistance of any other dispute resolution technique as long as it was not a further adjudication, which seems somewhat irrational. An alternative approach in some circumstances could lie within a careful wording of what further adjudications were about, so that the ongoing events were split into clearly identifiable and separate parts and constituted separate disputes. This would then prevent different binding decisions being made on the same issues or overlapping issues.

It will be apparent that sometimes it will be difficult to determine if a subsequent dispute is the same or different from an earlier one. The definitions of the disputes in the Notices of Adjudication may be crucial in determining this.

Paragraph 24 Enforcement of peremptory decisions

Section 42 of the Arbitration Act 1996 shall apply to this Scheme subject to the following modifications:

(a) in subsection (2) for the word 'tribunal' wherever it appears there shall be substituted in the word 'Adjudicator'

(b) in subparagraph (b) of subsection (2) for the words 'arbitral proceedings' there shall be substituted the word 'adjudication'.

(c) subparagraph (c) of subsection (2) shall be deleted, and

(d) subsection (3) shall be deleted.

Section 42, as amended, therefore states:

Enforcement of peremptory orders of tribunal

(1) Unless otherwise agreed by the parties, the court may make an order requiring a party to comply with a peremptory order made by the Adjudicator.

(2) An application for an order under this section may be made –

By the Adjudicator (upon notice of the parties),

By a party to the adjudication with the permission of the Adjudicator (and upon notice to the other parties).

(3) Deleted.

(4) No order shall be made under this section unless the court is satisfied that the person to whom the Adjudicator's order was directed has failed to comply with it within the time prescribed in the order or, if no time was prescribed, within a reasonable time.

(5) The leave of the court is required for any appeal from a decision of the court under this section.

Peremptory compliance has already been discussed in Paragraph 23 (1), where it is described how the courts considered that the more appropriate procedure to recover money would usually be the issue of court proceedings for the recovery of a debt and, where there is no money involved, to request a monetary injunction.

The decision peremptory enforcement provisions of the Scheme were initially one of its main areas of criticism. Similar comments were made in respect of enforcement provisions in other adjudication procedures. To some extent confusing the issue in this case was an underlying argument between lawyers as to whether or not all disputes (which could include a dispute on an adjudication decision) must finally be resolved by arbitration and not the courts when there were arbitration clauses in a contract.

In the context of Paragraph 24, problems arose because of the specific initial reference within it to the Arbitration Act. However, it is considered that the foregoing resulting wording is perfectly clear in its intent, i.e. that enforcement of an adjudication decision should be through the courts. If the wording had been quoted in full in Paragraph 24 instead of by reference to Arbitration Act amendments, it is suggested that there would have been no more problem with this than with provisions being proposed to achieve the same result in other adjudication procedures. A slight improvement might have been to add a simple statement such as 'For absolute clarity, enforcement of an Adjudicator's decision shall be through the courts and not through arbitration.'

In fact, despite the earlier criticisms, the courts (see various Selected Adjudication Summaries) have confirmed by their actions that they consider that they are an appropriate forum to resolve appeals against Scheme and other adjudication decisions.

Peremptory compliance and enforcement are discussed in more detail under those headings in Chapters 13 and 15 respectively.

Paragraph 25 Adjudicator fees and expenses

> *The Adjudicator shall be entitled to the payment of such reasonable amount as he may determine by way of fees and expenses reasonably incurred by him. The parties shall be jointly and severally liable for any sum which remains outstanding following the making of any determination on how the payment shall be apportioned.*

Whilst the foregoing is clear on Adjudicator fees and expenses, neither the Construction Act nor the Scheme makes express reference to the recovery of Parties' costs. The situation that applies to the current Scheme is that Party costs can be decided if both Parties requests them (i.e. Parties can agree to enhance the Scheme provided they do not detract from any express requirements of it or Construction Act). This would not be the case if one Party did not want costs awarded and said so. It will be noted from earlier comments under Paragraph 20, that this situation might change if currently proposed legislative amendments to the Scheme (at least for England) are carried through. In that case, an Adjudicator would be expressly prohibited from awarding costs.

It will be noted that several standard adjudication procedures (e.g. ICE, CIC, TeCSA) expressly exclude an Adjudicator from awarding costs, although it is permitted by the GC/Works Procedure.

This issue is discussed in more detail under the heading 'Allocation of costs and fees' in Chapter 13.

Paragraph 26 No liability except for bad faith

The Adjudicator shall not be liable for anything done or omitted in the discharge or purported discharge of his functions as Adjudicator unless the act or omission is in bad faith, and any employee or agent of the Adjudicator shall be similarly protected from liability.

Indemnity clauses and 'bad faith' have been discussed in detail under Section 108(4) in Chapter 2 dealing with the Construction Act.

Chapter 8

Possible Scheme Amendments

General

The Scheme (and/or possibly the Construction Act) may be amended by legislation. This Chapter describes the proposed amendments to the Scheme for England contained within the Draft of the Scheme for Construction Contracts (Amendment)(England) Regulations 2001 [hereafter the Draft Scheme Amendments]. Wales, Scotland and Northern Ireland are required to bring in their own changes. However, it must be noted that the following proposals are only in draft and might be amended or deleted entirely, and perhaps other may be added.

The currently proposed Draft Scheme Amendments.

The possible changes are as follows:

Paragraph 20 Amendment relating to costs

It is understood that the Government is relatively satisfied with how case law on costs has been resolved by the courts. However, if any changes are now made, the Draft Scheme Amendments indicate that provision could be as follows:

Regulation 2 (2) could add to Paragraph 20 of the existing Scheme, after the words 'connected with the dispute', the words 'but shall not take into account any matter relating to the legal or other costs of the parties arising out of or in connection with the adjudication'.

Thus, the first part of the revised Paragraph 20 could then read:

> *20. The Adjudicator shall decide the matters in dispute. He may take into account any other matters which the parties to the dispute agree should be within the scope of the adjudication or which are matters under the contract which he considers are necessarily connected with the dispute but shall not take into account any matter relating to the legal or other costs of the parties arising out of or in connection with the adjudication.*

Those consulted about proposed Scheme amendments took exception to the issue of costs in adjudication, it being considered that this distracted from ensuring that adjudication remained focused on the principal issue in dispute. Concern was heightened by a court decision (*Bridgeway* v *Tolent* [2000]) which supported a conditions of contract provision that a Party seeking adjudication had to pay all costs (including those of the Adjudicator and the other Party) whether it won or lost the ensuing adjudication.

The Government accepted that its policy is, or should be, for each Party in an adjudication to bear their own costs and this is why the amendment was proposed.

The proposal to incorporate it into in the Scheme and not the Construction Act means that the amendment would only apply to Scheme adjudications and the Parties could not change that provision. (It might, however, be possible in some circumstances for the Parties to recover their adjudication costs by court action.) Contractual adjudication provisions, on the other hand, could still make provision for costs, e.g. even in a contract which is Construction Act-compliant it could validly provide for costs and require the Adjudicator to decide and rule upon those costs.

It is, perhaps, interesting to note that a costs exclusion in Scheme adjudication would apply to legal or other costs arising not just 'out of' the adjudication but also 'in connection with' it.

Paragraph 22 Amendment relating reasons

Existing paragraph 22 could be deleted from the existing Scheme and a new Paragraph 22 substituted. This could read:

> *22 (1). Subject to paragraph* [22] *(2), if requested by one of the parties to the dispute, the Adjudicator shall provide reasons for his decision.*
>
> *22 (2). The Adjudicator may set a deadline for the parties to request the reasons for his decision and where a request is submitted outside that deadline, the Adjudicator shall not be required to comply with paragraph* [22] *(1).*

The current Scheme has no time limit related to the request for reasons and, in theory, a request for reasons could be made at any time (even very shortly) before, or even some time after, the decision. It was considered by some commentators that the approach of some Adjudicators might be different depending on whether they knew whether they had to provide reasons or not, and others thought it impractical for an Adjudicator to be asked for reasons some time after issuing its decision.

The proposed Paragraph 22 (2) amendment would allow Adjudicators at any time to set deadlines by which the Parties might request reasons. It will be noted that if the Adjudicator did not do so, Paragraph 22 (1) would still apply and a Party could request reasons at any time.

Paragraph 22A Amendment relating to correcting errors

It is understood that this is another area where the government is now relatively satisfied with how case law has developed. However, the Draft Scheme Amendments suggest that a new Paragraph 22A could be added to the existing Scheme which could read:

> *22A (1). The Adjudicator may, on his own initiative or on the application of a party, correct his decision so as to remove any clerical mistake or error arising from an accidental slip or omission.*
>
> *22A (2). Any application for the exercise of the Adjudicator's powers under paragraph 22A (1) shall be made within 5 days of the date that the decision is delivered to the parties or such shorter period as the Adjudicator may specify in his decision.*
>
> *22A (3). Any correction of a decision shall be made as soon as possible after the date that the application was received by the Adjudicator or, where the correction is made by the Adjudicator on his own initiative as soon as possible after he becomes aware of the need to make a correction.*

When adjudication was first introduced, it was suggested that adjudication procedures and especially tight deadlines would result in mistakes being made in decisions. This has happened. Unfortunately, the current Scheme makes no provision for a 'slip rule' and some Adjudicators believe that they could not correct their own recognised errors.

It was also subsequently made clear in a number of court decisions (e.g. see *Bouygues* v *Dahl-Jensen* [2000] Appeal Court) that such mistakes of fact would not invalidate the temporary binding nature of an Adjudicator's decision. The resultant position then became confused, with some Adjudicators apparently believing that they had no power to amend their decision and other Adjudicators importing either by agreement or at common law a form of the 'slip rule'. The latter approach was endorsed in a number of later adjudication decisions.

To formalise the situation, at least for statutory adjudications, the draft proposal new Paragraph 22A incorporates a means of correcting slips into the Scheme. This is not open-ended but, as can be seen from the Paragraph, is carefully constrained as regards both the nature of correctable errors and the time within which they may be corrected.

Once again, the incorporation of the amendment in the Scheme and not the Construction Act would mean that the amendment to correct slips would only apply to Scheme adjudications. Contractual adjudication provisions on the other hand could still exclude provision for slips, i.e. even if a contract was Construction Act-compliant it could validly omit such a provision.

Part II

The conduct of an adjudication

Chapter 9

Adjudicator appointments

Introduction

The appointment of an Adjudicator may be by being named in the project contract documents, by being agreed post-contract before a dispute arises, or by being appointed after a dispute has occurred. The appointment may be made either by the Parties, by a nominating person or body named in the conditions of contract, or in the Scheme by an Adjudicator Nominating Body, as discussed elsewhere in this text. The options available will depend on the attitudes of the Parties, the pertaining adjudication procedures and the conditions of contract. The latter should be referred to for specific procedures and wordings in each case that arises.

A common feature of whichever method is followed is that after a dispute arises, a potential Adjudicator, even one pre-named in a contract, should always have an opportunity to assess its suitability and confirm availability and willingness to act, before an appointment can be confirmed.

Factors to be considered before an Adjudicator agrees to an appointment

At the time of being approached, a potential Adjudicator should have been provided with sufficient preliminary information, including the title (if not a copy) of the pertaining adjudication procedure, on which to decide whether to act or not. The potential Adjudicator normally has a strict time limit (which is the first information the Adjudicator should determine from the adjudication procedure), typically 2 days, by which to respond.

It is suggested that the potential Adjudicator, as far as it is possible at this stage, should consider jurisdictional issues (see Chapter 10 Jurisdiction), as well as such other matters as its terms and conditions of engagement. In fact, the Courts have not yet made clear the extent to which, under adjudication, an Adjudicator *must* take cognisance of many of these issues, or should do so only when an objection has been raised by one of the Parties. This does not, however, prevent an Adjudicator from considering such matters, not least when it is possibly in the Adjudicator's own interest to do so. Issues include:

- If the contract came into force before or after the Construction Act became applicable, i.e. 1 May 1998. If before, this does not exclude adjudication, but contractually only the adjudication procedures, if any, in (or incorporated by reference in) the construction contract will apply. If later, the Act will, and the Scheme may also, be relevant.

- If the contract from which the dispute arose appears to be one that comes under the Construction Act, e.g. it involves a construction contract as defined in the Act, and is not a type of operation that is excluded. If it is excluded, this still means that the adjudication procedures, if any, in the construction contract will apply. Standard adjudication procedures will usually be Construction Act-compliant anyway.
- Whether the Referring Party is opting to use (a) any stated contractual adjudication procedure, or (b) if it considers the latter to be non-Construction Act-compliant and is opting to use the Scheme instead. As has been noted elsewhere in this text, it is only the Referring Party that has the option to do this under the Construction Act, not the Adjudicator.
- If the Adjudicator's proposed appointment appears to comply with any appointment requirements of Act, the pertaining adjudication procedure or rules, and any construction contract adjudication clause requirements.
- If the Adjudicator has a potential conflict of interest by any relationship with the Parties, financial or otherwise, and action to be taken if so. Normally the Adjudicator, as a minimum, will be expected to advise both Parties if such a conflict is perceptible.
- That the Adjudicator has any problems associated with any specified adjudication procedures, e.g. that they lie within its knowledge and/or experience, or whether they appear to be complex, such as ad hoc or the GC/Works/1 requirements.
- That the Adjudicator is available to meet the timetable for a decision.
- That the dispute is of a type the Adjudicator is prepared to adjudicate on, e.g. the area of expertise required, whether technical, quantum, timing or contract interpretation-related, and so on.
- If the adjudication appears particularly complicated.
- Whether the dispute is likely to be of a manageable size for the timescale (see *Balfour Beatty* v *Lambeth* [2002]).
- Whether any general assistance might be required from others under the control of the Adjudicator on document management, research, technical or quantum preparation and/or analysis etc. (notwithstanding that the Adjudicator remains responsible for that work), which would assist achieve the timetable, produce a better informed decision and probably be at lower fee rates than if the Adjudicator had undertaken the work itself. If so, what procedures need to be instigated to obtain any approval considered necessary and ensure that any additional fee rates are notified to the Parties.
- Whether the procedures or rules require the Adjudicator to be 'a natural person acting in his personal capacity' or similar, and whether that creates any problems that need notifying and resolving.
- How flexible one or both of the Parties might be regarding timescale.
- If the dispute can be split into simpler parts. As ideally this should have been done before the Adjudicator had been approached (e.g. within the original notification of the intention to refer a dispute to adjudication), it might be contractually difficult to reorganise after the appointment of the Adjudicator.

- Whether it is likely to involve joining with another dispute on the same subject and any potential related difficulties (e.g. timescale, whether different procedures or rules might apply).
- Whether any earlier dispute on the same subject had had a previous adjudication, conciliation, expert determination, or other ADR decision and what the relevant procedures or rules might say on the matter (i.e. some, like the Scheme, exclude an Adjudicator proceeding in certain circumstances).
- Whether at any earlier adjudication, ADR, or conclusion of talks between the Parties, a decision was agreed to be (or had become by default, e.g. ICE Procedure) a conclusive 'final and binding' agreement between the Parties, any grounds provided for reopening it, what the relevant adjudication procedures or rules might say about such matters, and whether the Adjudicator might want, or suggest that the Parties obtain, legal advice before proceeding further with the appointment.
- Whether a 'dispute' had actually arisen, e.g. had the issue being claimed been presented to the other Party and had any response been disagreed with, had there been no response etc. (See comments on disputes in Chapter 14.)
- Is the issue one arising 'under' or 'in connection with' the contract and do the adjudication procedures confirm that whichever one of those it is can be dealt with.
- Whether there are court clarifications required on jurisdiction or other matters that might be best dealt with in advance of proceeding further. (See *Palmers* v *ABB Power* [1999].)
- Are there likely to be problems to be sorted out on the Adjudicator's Agreement, e.g. is a compulsory pro forma Agreement acceptable to the Adjudicator? Does it, or the adjudication procedures, or the contract, include for any Advance payment or other means of security for fees and expenses, such as withholding the decision until outstanding amounts had been paid? (See later in this chapter for issues to be considered in an Agreement.)

Based on the foregoing, which might require some discussion with the Parties, an Adjudicator should be able to make a decision on how to proceed. The options are:

Don't accept appointment

It will be clear from the foregoing that there might be perfectly valid reasons other than limited time availability and lack of relevant experience, for not accepting an appointment. There may also be an element of cherry-picking for those less experienced Adjudicators, or for whom the potential difficulties of a complicated adjudication are not worth having. However, as more experience, confidence and familiarity with Construction Act and procedures develop, as well as knowledge gained from reports and discussions in the technical press and at seminars, and also court decisions on the interpretations of areas of the Act and procedures, the more that element should reduce. The first option for an Adjudicator, however, remains to refuse an appointment.

Accept appointment

If a potential Adjudicator decides that an adjudication appointment can be accepted, consideration of issues such as those listed will assist determine what terms and conditions are important, if not essential, to have included in the Adjudicator's Agreement, assuming that it is one where there is scope for negotiation.

Once appointed, an Adjudicator must then do its best in the circumstances and reach a decision in the agreed timescale. Most disputes referred to adjudication are, in fact, of a relatively simple nature, so should not constitute a major problem for someone with the right experience and adjudication training.

In addition, although acting as an individual, additional assistance such as the following will usually be available should an adjudication start proving more difficult or time-consuming to sort out than initially anticipated:

- Independent Adjudicators will usually have copies of published literature, case reports, seminar notes, other practitioners or ex-colleagues who can provide a sounding board on what to do in difficult disputes.
- An Adjudicator is not normally expressly excluded from also being a company employee, director or partner. A significant number of Adjudicators will therefore have ready access to, and (particularly in complex multifaceted disputes) may need to rely on, ad hoc advice and possible assistance from in-house colleagues and other members of staff, e.g. listed Adjudicators, contract specialists, technical experts in the relevant area of a dispute or quantum specialists.
- As stated earlier, an Adjudicator might also use more general in-house support services, under the control and responsibility of the Adjudicator, for document management and technical and quantum assistance. If initial approval had not been obtained from the Parties to do this, it might be possible to obtain their agreement during the adjudication itself.
- Adjudication procedures usually will permit an Adjudicator to obtain outside technical and legal advice without obtaining the approval of the Parties, although the latter should be advised that this is taking place.

Review the situation as the adjudication proceeds

An Adjudicator's knowledge of a dispute will increase as an adjudication proceeds. Indeed, one or more of the items listed when considering whether to accept the appointment initially might be the subject of later alternative submissions by the Responding Party. Such issues must be reviewed by the Adjudicator, e.g. the Adjudicator might consider in retrospect that a 'dispute' had not, in fact, arisen, after hearing from both Parties.

An Adjudicator should also recognise when a situation has developed significantly, or become more complex, than originally advised when accepting the appointment, and must review whether it is still possible to provide a proper decision at all, or a proper decision in the available timescale (see *Balfour Beatty* v *Lambeth* [2002]). Also of relevance is whether it is possible for the Parties to continue to have a reasonable opportunity to present their cases and deal with those of the other Party, i.e. to comply with the Rules of Natural Justice (see Chapter 11).

Other than considering matters already mentioned (such as obtaining in-house or outside advice and assistance, attempting to agree a time extension, and so on), there would appear to be two further and concluding options available:

- Provide a 'decision' that there is no dispute, or that the dispute is, or has become unsuitable for adjudication, or unsuitable within the available timescale and the requirements of natural justice, etc., as appropriate, letting the Parties decide if they want to appeal that decision. This option has potential problems, probably based on whether a true decision has, in fact, been made, including a possible refusal to pay an Adjudicator's fees, a request for the return of money already paid, or even a claim for abortive costs expended by a Party up until that point.
- Resign, using contractual resignation provisions, which may allow an Adjudicator to be still reimbursed fees and expenses reasonably incurred to date. At worst those fees and expenses would be forfeited.

Urgency of Appointment

Some adjudication procedures may not need to be Construction Act-compliant, e.g. the contract concerned may not be a 'construction contract' or a Referring Party does not want to take up the option of statutory adjudication as it prefers to use the contractual adjudication procedures (which may or may not be Construction Act-compliant). Such a contract could, for example, provide that adjudication can only take place after a certain event occurs or that the appointment of an Adjudicator and the referral of the dispute to that Adjudicator should take place over a longer period than is required by the Construction Act , for example, 15 days. In general, and depending on the exact wording, provided the Parties are using their best endeavours to comply with that timetable, then a slippage of a day or so in achieving it may not be fatal in relation to the appointment of, and the referral of the dispute to, the Adjudicator.

For statutory Scheme adjudications, the requirements of the Scheme and the minimum requirements of Section 108 of the Act must be followed. Section 108 (2)(b) is one of the essential requirements of the Construction Act, stating that adjudication procedures shall provide a timetable with the object of securing the appointment of the Adjudicator and the referral of the dispute to the Adjudicator within 7 days of the notice of an intention to refer a dispute to adjudication. As discussed in the commentary to Section 108 (2)(b) in Chapter 2, it could be interpreted that the 7-day requirement is mandatory. This is the position taken by Paragraph 7(i) of the Scheme in Chapter 7, where 7 days is a stated express requirement which could be argued reinforces the 'mandatory' interpretation of the Act. If so, then any failure to achieve both the appointment of, and the referral of the dispute to, the Adjudicator within 7 days might affect the validity of any eventual decision. If the Section is regarded as merely directory, however, then some slippage in the achieving of both the appointment of the Adjudicator and the referral of the dispute to the Adjudicator within the 7 days of the Notice of Adjudication may not be fatal.

Adjudicator Agreements

A potential Adjudicator's acceptance will normally be linked to the terms and conditions of its appointment.

As indicated, there will probably be a stated limited mandatory or indicative time scale for either the appointment and the referral of a dispute to the Adjudicator, either for these two activities combined or, in some adjudication procedures, for these as separate events. Consideration of terms and conditions will therefore normally need to run in parallel with the appointment process.

It will be noted that Adjudicator Agreements are not mentioned in either the Construction Act or the Scheme and there is no strict statutory requirement for the Parties to an adjudication to have such an Agreement. A Scheme appointment can therefore be made by means of a simple letter to that effect.

Many construction industry disputes arise because there is just an exchange of letters and no, or no adequate, contract in place between the Parties involved. Where work commences without a contract, the service provider remains entitled to payment on a *quantum meruit* (i.e. reasonable remuneration) basis. Similarly, if there is no formal Adjudicator's Agreement, an Adjudicator can still commence in the expectation that at the very least reasonable fees and expenses will be paid, and that this will be supported by the courts.

However, most professionals engaged as Adjudicators will not want to become embroiled in court debt recovery proceedings with their associated loss in potential fee earning time and aggravation. Wherever possible, it is therefore always prudent, and most Adjudicators prefer, to have a written contract or services agreement, and one that covers all reasonably foreseeable circumstances. A major issue for a potential Adjudicator will thus normally be the rapid agreement of the terms and conditions of its appointment. As discussed elsewhere in this text, these should include obtaining indemnities from the Parties against contractual and third party liabilities (including any that might arise from the Contracts (Rights of Third Parties) Act 1999), if those have not already been adequately provided for within the adjudication provisions or underlying conditions of contract. The timetable may mean that an Adjudicator accepts an appointment in principle, subject to sorting out terms and conditions later. It will be noted that an institutional appointment normally becomes effective from the date the institution notifies the Parties of the name of an Adjudicator who has agreed with the institution that it is willing and able to act.

Several organisations and most institutions that make appointments publish their own adjudication procedures or rules and also have their own pro forma Adjudicator's Agreements, which can be either sample forms for guidance on what is required, or are, more commonly, compulsory. Pro forma sample forms can assist obtain consistency between what is contained within an adjudication procedure and what is in a final amended Adjudicator's Agreement. Compulsory pro forma Agreements can, however, limit scope for individual Adjudicator input, except, usually, on hourly rates. This is unfortunate, because it is considered that some of the pro forma Adjudicator Agreements can fall significantly short of what most professionals might want and expect to have included for their own protection within their usual professional services appointments.

Notwithstanding any apparent limitations inherent in a compulsory pro forma Agreement, as noted in the chapters dealing with particular adjudication procedures, there is nothing to stop an Adjudicator from proposing additional terms and conditions, or even variations to such Agreements. Equally, the Parties are not obliged to accept the Adjudicator's proposals. If they do not accept them, the Adjudicator can either cease having any further involvement (with whatever consequences that brings) or else proceed as if it hadn't made any proposals in the first place.

Sometimes issues relating to an Adjudicator's terms and conditions are included within the adjudication procedures or even in the underlying conditions of contract, without necessarily being stated in a pro-forma agreement. An example might be the indemnity clauses mentioned earlier (although, as described elsewhere in this text, they are sometimes not totally satisfactory regarding the extent of cover they provide). These will therefore be in place even if an Adjudicator decides to proceed without any, or any totally satisfactory, Agreement.

Signatures on Agreements

Adjudicators pre-named in contracts will invariably have agreed terms and conditions in advance, or will have committed themselves to accepting a pro forma Agreement. For their part, by signing the underlying construction contract, the Parties will probably also be deemed to have agreed to any included pre-named Adjudicator and to any pre-stated terms and conditions without the need of a further set of signatures to be appended to an Adjudicator Agreement later, if a dispute arises.

Where there is no pre-named Adjudicator, it is assumed that to have got to the stage where an Adjudicator's Agreement is about to be finalised (no matter how the nomination has arisen), at least one of the Parties will be in agreement with the appointment, otherwise it is probable that the Adjudicator will be wasting its time if neither Party wants to cooperate. It would therefore seem likely that, at the least, there would always be the signature of one Party and the Adjudicator itself on an Adjudicator's Agreement, and that a contract would exist between the two of them. It could be argued that there should then be no problem to the Adjudicator getting its fees paid by the contracting signing Party. However, it should be noted that it has been argued that arbitration is a trilateral agreement between the Parties and the Arbitrator, and that an Arbitrator cannot deal with only one Party for personal benefit or enter into any fee arrangement with one Party that the other Party objects to. It is possible that this view could equally be applied to adjudication.

Compulsory pro forma Agreements invariably require both Parties to sign them within a certain number of days. Adjudication procedures may have express fall-back provisions (e.g. the JCT Procedure, Paragraph 5.6) allowing an Adjudicator to proceed to a decision even in the case of non-signature, which is likely if one Party decides it wants nothing to do with the adjudication and decides not to cooperate.

Usually, however, there is no stated fall-back position. Of possible relevance is the Court of Appeal decision (*Andrews* v *Bradshaw and Randell*

[1999]) that where an Arbitrator accepted an appointment from a professional body, the Arbitrator (a) was not entitled to insist upon the Parties using the Arbitrator's own terms and conditions and (b) should not enter into an agreement with only one of the Parties when the other Party refused to do so. The courts might well consider that the same applied to adjudication.

The practical options then are for the Adjudicator to either resign once the time limit for signing has elapsed, or else to proceed anyway on the understanding that it should (at least) get a *quantum meruit* reasonable remuneration judgment from the courts, if it considered it worthwhile spending more money and time chasing any outstanding debt.

The foregoing begs the question as to why an Agreement needs signatures with their inherent problems anyway, if either a fall-back position or a *quantum meruit* entitlement will ensure fee reimbursement whether there are signatures or not.

An alternative approach, appropriate to some situations where there is no compulsory pro forma Agreement, might be the common one adopted in practice by professionals when receiving instructions on a project requiring rapid action. In that case, the professional confirms by return to the instructing Party the instructions received, its willingness to act, encloses what it considers to be reasonable terms and conditions to be applicable to the appointment, and states that it is proceeding forthwith with the assignment. If there is no objection from the contracting Party(ies) within a reasonable time (which will mean a very quick response in the case of a short timescale) and the Party(ies) communicates with the professional for a time as if the terms and conditions were acceptable, the courts will probably support those terms and conditions. If immediate objections are raised about terms and conditions they will hopefully only be on one or two points, so the professional would mostly get what it wanted anyway, whereas if it had not made the proposals initially it would not have got them. It is considered that wherever possible, something similar to the foregoing will be the better approach for an Adjudicator.

Amendment of standard pro forma Adjudicator Agreements

Pre-named Adjudicators are usually those named and agreed to by the Parties within construction conditions of contract at the time of tender award.

For a pre-named Adjudicator, there is thus usually an opportunity before issue of tender documents to discuss and agree any amendments to a pro forma Adjudicator's Agreement included within the tender documents. Any pre-contract award Agreement amendments are no different in principle from a client adjusting standard conditions of contract before the latter go out to tender, a procedure that is probably more the norm than using unaltered standard conditions of contract. It requires care and the use of staff or advisors with relevant experience.

Similarly, for Adjudicator Agreements care needs to be taken so that there is consistency between an Agreement, the underlying adjudication procedures and any relevant clauses in the construction contract. It is sensible to state a clear order of document precedence anyway.

For a post-contract appointment, where there is an obligatory pro forma Adjudicator's Agreement and no possibility to alter it under the contract, the situation is different. A proposed Adjudicator may demand significant amendments (i.e. for its terms and conditions to take precedence over those in the pro forma Agreement, those that might also be relevant within the underlying adjudication procedures and/or those in the conditions of contract). There might be some delay before the Parties found out the seriousness of this situation and then further delays before another Adjudicator could be appointed.

If a potential Adjudicator was someone that was particularly wanted by either or both Parties (say, because of particular reputation, Adjudicator experience or specialised technical knowledge), there seems no way around an intractable problem that, it will be noted, has only arisen as a result of a requirement of an adjudication procedure (not the Act) making a pro forma Agreement compulsory. It is suggested that such a requirement must be of questionable overriding merit.

Issues that might be included in an Adjudicator's Agreement

In situations where an Adjudicator will only accept an appointment on its own terms and conditions, there are a number of issues that the Adjudicator might want included, or spelt out clearly, in pertaining circumstance and/or for different degrees of client sophistication. These singly, in combination or as alternatives are included in the following check list, many of which have already been discussed earlier in the text.

- An advance appointment fee, to be paid by one or both Parties to the Adjudicator within a stated time limit, otherwise an Adjudicator can resign and be entitled to be paid fees and expenses incurred to that date. As amounts are likely to be relatively small, it is probably impractical to make any provision for any interest that might be earned over the short duration of an adjudication.
- An entitlement to draw invoiced amounts from that advance fee.
- Any other preferred means of security for fees and expenses. For example, a requirement for payment of all the Adjudicator's fees and expenses before it needs to release its decision.
- A fixed lump sum for the whole adjudication, based, for example, on the estimated time it might take, and when to be paid, in whole or in part, by one or both Parties.
- A fixed lump sum for the whole adjudication based on the amount in dispute, and when to be paid, in whole or in part, by one or both Parties.
- An hourly rate.
- Perhaps a daily rate for each whole day away from a stated 'usual office' location.
- Travel and associated unavoidable time to be included, probably at the regular hourly rate to reflect the lost potential alternative earning time and the likelihood that some of that time might be actually be spent productively on work associated with the adjudication.
- First class rail travel.

- Business class air travel.
- VAT to be included or excluded.
- Terms and conditions that would apply to staff under the control of, and who provide any day-to-day in-house assistance to, the Adjudicator. These may be stated as either specific rates (e.g. 'quantity surveyors will be charged out at £X per hour') or rates from within a range (e.g. 'quantity surveyors will be charged at a rate within the range of £Y to £Z per hour').
- Procedures for the instruction and reimbursement of persons providing legal and technical advice, on the understanding that the Parties would be advised before reference was made to those persons.
- Whether contractually reasons will be provided or are excluded, or whether that is an option that can be instructed by one or both Parties and the latest time for that instruction to be given (e.g. 14 days before the date due for notifying the adjudication decision).
- That the fees and expenses incurred in preparing reasons will be charged unilaterally to any single Party requesting reasons (where such a request is permitted by the adjudication procedure).
- Fees and expenses otherwise to be split equally between the Parties.
- Fees and expenses otherwise to be borne in total by the Referring Party.
- Circumstances where an Adjudicator *must* resign and be paid fees and expenses (e.g. under the Scheme an Adjudicator must resign where it becomes apparent that the dispute is the same or substantially the same as one already previously decided upon by an Adjudicator and the procedure precludes the later Adjudicator from dealing with it).
- Circumstances where an Adjudicator *could* resign and be paid fees and expenses:
  - In the opinion of the Adjudicator the dispute varies significantly from that originally referred for confirmation of willingness to act and the Adjudicator now considers that it is not now competent or able to decide it.
  - A full appointment fee has not been paid within the proscribed time.
  - The Adjudicator becomes aware of any interest, financial or otherwise, in any matter relating to the dispute which the Adjudicator considers may cast reasonable doubt on the impartiality of its decision.
  - The combining of two or more disputes leads to effective revocation of the Adjudicator's appointment due to another Adjudicator now dealing with the dispute.
  - The Adjudicator becomes aware after its appointment that the Referring Party has not followed correctly all pre-dispute contract procedures either in terms of establishing that an adjudicable dispute exists, or in the selection of the Adjudicator.
  - After appointment, the timetable for the decision is required to be extended by one or both of the Parties and the Adjudicator, in its opinion, is unable to comply with those amended timetable requirements.

- Jurisdictional situations where an Adjudicator should be paid fees and expenses:
  - There is a dispute over the Adjudicator's jurisdiction, with the Referring Party considering that jurisdiction exists and wishing the adjudication to proceed, it being decided by arbitration or the courts some time later that there is no jurisdiction. In this case, the Referring Party could be made solely responsible for paying the Adjudicator's fees and expenses.
  - Both Parties agree after the adjudication commences that doubt has arisen over the Adjudicator's jurisdiction, and it is decided by arbitration or the courts some time later that there is no jurisdiction. In this case, both Parties remain equally responsible for paying the Adjudicator's fees and expenses.
  - Both Parties give the Adjudicator the power to decide a matter of jurisdiction, but that decision is later overturned by arbitration or the courts. In this case, both Parties remain equally responsible for paying the Adjudicator's fees and expenses.
- Circumstances where an Adjudicator's appointment could be revoked by the Parties but the Adjudicator is entitled to be paid fees and expenses:
  - The Parties settle outside the adjudication or have the Adjudicator produce a pre-decision consent award.
  - For lost future time allocated to that adjudication (e.g. cancellation fees should the Parties settle at short notice outside the adjudication, leaving the Adjudicator with a gap in allocated time that cannot be filled. If the amount or method of calculation of the cancellation fee is clearly stated, the Adjudicator should be more easily able to deduct the amount from its advance appointment fee or other means of security).
- Circumstances where an Adjudicator can resign but not be entitled to be paid fees and expenses.
- Circumstances where an Adjudicator's Agreement could be revoked by the Parties but the Adjudicator is not entitled to be paid fees and expenses, e.g. where services are terminated due to bad faith or excess of power on the part of the Adjudicator.
- Joint and several liability for the payment of fees and expenses.
- Provisions for payment of fees and expenses in instalments for disputes where an extension, or extensions, of time beyond 28 days for the Adjudicator's decision have been agreed.
- Liability for specified rates of interest on payments outstanding [x] days after invoicing by the Adjudicator.
- The necessity for a full indemnity, including specific reference to both negligence and to cover against any liabilities that might arise under the Construction (Rights of Third Parties) Act 1999, to be provided by the Parties to the Adjudicator.
- A statement that the Adjudicator's obligation is to exercise reasonable skill and care only (to permit any residual risk to be covered by more general professional indemnity insurance).

- What procedures will be adopted, or the options that will rest with the Adjudicator, if an Adjudicator's Agreement includes a requirement that both Parties sign and date it within a certain period, one or more does not do so, and there is no express fall-back position within the adjudication procedures. Options could include either to resign with entitlement to payment of fees and expenses, or that the Adjudicator would record the fact to both Parties and then proceed with the adjudication anyway in the expectation of at least *quantum meruit* reimbursement.
- Where one Party pays the whole amount of an Adjudicator's fees and expenses, some or all of which should have been paid by the other Party, an appropriate adjustment will be made by the Adjudicator to the calculation of the final full fees and expenses reimbursable by each of the Parties.
- The need for, and extent of, confidentiality.

It is further suggested that:

- Where adjudication procedures include Adjudicator terms and conditions that an Adjudicator agrees with, exactly the same wording should be used in the Adjudicator's Agreement for the same topics, to avoid potential confusion.
- To assist avoid confusion, have a statement included in the Adjudicator's Agreement that the Adjudicator's current terms and conditions take precedence over any stated in the contract or adjudication procedures.

It will be evident that only some of the foregoing would be required, as some are alternatives to others, some might not be considered too important, and there can be a degree of combining and simplification. However, none is considered to be unreasonable, and if Parties believed them to be so, this should be a warning to an Adjudicator and provide all the more reason for having them included in the Agreement.

Appendix 1 contains an example Adjudication Agreement that incorporates wording covering many of the foregoing listed options.

Chapter 10

Jurisdiction

Introduction

To be able to deal with an adjudication, there must be a dispute. That itself can be a contentious issue (see Chapter 14). An Adjudicator must have the relevant authority to deal with that dispute, i.e. what the Adjudicator does must fall within its jurisdiction. That authority comes from the underlying conditions of contract and/or the pertaining adjudication procedures.

A Party with a dispute will probably first look to the dispute resolution procedures contained within its contact with the other Party in order to resolve that dispute. Those dispute resolution procedures will nowadays often, but not always, include adjudication. The Party with the dispute must comply with whatever procedures are stated in the contract and its adjudication provisions in order to appoint an Adjudicator, and the Adjudicator, once appointed, must also proceed in accordance with the stated procedures to reach a decision. Alternatively, a Party with a dispute on a contract that comes under the Construction Act has an option to use an adjudication procedure complying with the 8 minimum requirements of the Act, all as discussed earlier in this text. It may be that the contractual adjudication provisions already comply with those requirements and so once again those contractual procedures must be used for the Adjudicator appointment and the subsequent actions of the appointed Adjudicator to reach its decision.

Thus, in both the foregoing circumstances, for the Adjudicator to be judged as acting within its jurisdiction, the appointment procedure and subsequent actions of an Adjudicator must be in accordance with the contractual adjudication provisions.

It is only when a Referring Party wishes to use statutory Scheme adjudication that other problems arise. Firstly an entitlement to do so must be established, eg is the contract one for work that falls under the Construction Act? Is it in writing? etc. If these are not met, a Referring Party does not have an option to use the Scheme (unless it is incorporated as a contractual adjudication procedure) and any Adjudicator appointment procedure and subsequent actions of the Adjudicator to resolve a dispute using Scheme procedures would be invalid due to a lack of jurisdiction.

The Notice of Adjudication served upon an Adjudicator at the outset of the adjudication process will provide an Adjudicator with some impression of what the referred dispute is about. In some instances an Adjudicator, in the early stages of the adjudication, might well be in touch with the Parties anyway, thus enabling issues to be better understood. At this point an Adjudicator should be able to assess more fully any preliminary, including jurisdictional, matters which require to be formally addressed.

If a Party has raised a jurisdictional matter then the Adjudicator has to deal with it, but the extent to which an Adjudicator is under an obligation to raise any jurisdictional problem not yet pointed out by a Party is, as yet, untested by the courts.

Some Adjudicators will initially attempt to address jurisdictional issues by telephone, with all such conversations being properly and timeously recorded in writing to the Parties. However, a number of cases (e.g. *Woods Hardwick* v *Chiltern* [2000]) have indicated that there are dangers inherent in such an approach. There may also be problems with attempting to resolve jurisdictional matters solely by correspondence, as anything received or sent to one Party must be copied to the other Party in order to give them a chance to respond, and exchanges between them and the Adjudicator, if not tightly controlled, might take an unacceptably long time. Another approach may be to hold a preliminary meeting at which any jurisdictional issues can be fully discussed between the Adjudicator and both Parties. A preliminary meeting will anyway be an opportunity for the Adjudicator to give its instructions regarding responses, time limits, find out whether reasons are required etc.

Whichever approach is used, the Parties will sometimes agree, with appropriate extensions of time, that the adjudication be postponed until any complex jurisdictional objections have been decided by the courts.

There can be significant differences between the jurisdictional powers given in contractual, compared with statutory Scheme adjudications. It will also be noted that the Adjudicator's powers and duties are currently more circumscribed, both within the Scheme and by court rulings related to the Scheme, than is the case for contractual adjudications.

Jurisdictional provisions of contractual adjudication procedures

Although many now do so, contractual adjudication procedures do not necessarily need to comply with any of the statutory requirements of the Construction Act, only with what is required by the pertaining conditions of contract themselves. Such contracts could provide that an Adjudicator is entitled to open certificates or previous adjudication decisions declared to be final and that an Adjudicator is not obliged to resign if it is found that the dispute is one that has been previously decided by another Adjudicator. An Adjudicator can thus be given judicial authority by the Parties of much wider scope than is in the statutory Scheme.

Contractual adjudication procedures that wish to eliminate an option for a Party to choose the statutory Scheme procedures are required to comply only with the eight essential minimum requirements of the Construction Act described earlier in the text. Provided that the contractual adjudication procedures do so, they are otherwise free to include any additional or enhanced adjudication provision they want. For example, Construction Act-compliant contractual adjudications, unlike statutory Scheme adjudications, can additionally provide that any decision upon jurisdiction by the Adjudicator is to be binding upon the Parties. In one case, the courts (*Nolan Davis* v *Catton* [2000]) held that in their contract the Parties had agreed to be bound (albeit on a temporary interim basis) by the Adjudicator's decision on its jurisdiction. This judicial authority is again of wider scope than permitted by the Scheme (but is not excluded by the Act).

Jurisdictional provisions of the statutory Scheme adjudication procedures

Neither the Construction Act nor the Scheme (with limited exceptions (see 'Jurisdiction given in adjudication clauses', later in this chapter)) give an Adjudicator the power to determine its own jurisdiction. Accordingly, common law applies. In England the common law position is that an Adjudicator is not entitled to decide upon its jurisdiction in a manner which binds the Parties. That is not to say that an Adjudicator cannot consider its jurisdiction. It simply means that any decision which the Adjudicator comes to is not, in law, regarded as binding upon the Parties and therefore may form a defence to enforcement proceedings. It has been suggested by the courts (e.g. *Christiani & Neilsen* v *The Lowry Centre* [2000]) that even though not binding, it would be sensible and even desirable for an Adjudicator to consider and rule upon its own jurisdiction in the first instance.

Some typical jurisdictional issues

Amongst other matters, Adjudicators, in the first instance, may need to consider the following jurisdictional issues:

1. Is there a bone fide 'dispute'?
2. Was the Adjudicator appointment process valid?
3. Is the contract a 'construction contract'?
4. Are the operations 'construction operations'?
5. Is the contract 'in writing'?
6. Do the contractual adjudication procedures comply with the eight minimum requirements of the Construction Act?

It will be noted that items 3 to 6 apply only were the option to choose the statutory Scheme has been selected.

Occasionally Referring Parties do not properly consider such issues when issuing a Notice of Adjudication and, even if they do so, Responding Parties may not agree, with the result that jurisdictional matters arise which an Adjudicator may need to address, or have decided, immediately before proceeding further with an adjudication.

The Parties should be given an opportunity by the Adjudicator to address any jurisdictional issue and to respond to each other's points. Having considered all submissions and advice given, the Adjudicator can then decide the jurisdictional point. That decision may, of course, be revisited by the courts if required by one or both of the Parties, but meanwhile the direction of that particular adjudication will have been determined.

It will be noted that some issues are not decisions which the Adjudicator is entitled to take, or take alone. For example, the alternative of statutory Scheme adjudication is merely an entitlement of the Referring Party and not an obligation that can be unilaterally imposed by an Adjudicator. This is discussed in more detail in the commentary following Section 108 (5) in Chapter 2 dealing with the Construction Act.

As has been described elsewhere in this text, an Adjudicator has no jurisdiction, *inter alia*, where:

(a) There is no 'dispute', or in a potential statutory Scheme adjudication there is no 'construction contract' as defined by the Act, and/or
(b) The Adjudicator goes outside its terms of reference, i.e. it has exercised an 'excess of jurisdiction' (say by deciding an issue that was not referred to it).

The courts (*Carter* v *Nuttall* [2000]) would possibly call (a) a 'threshold' jurisdictional matter, that goes to whether an adjudication should even be commenced in the first place, and (b) an 'internal' jurisdictional issue, which can be raised during any enforcement proceedings.

Where there is a doubt as to whether there is a 'construction contract', it can be ruled upon by an Adjudicator, but must ultimately be decided by the courts (*Sherwood & Casson* v *Mackenzie* [1999], *Workplace Technologies* v *E Squared and Anr* [2000] and *Butler* v *Merewood Homes* [2000]).

As long as an Adjudicator is acting within its jurisdiction, the courts will support the Adjudicator's decision even if it has made factual legal or procedural errors. The courts will support decisions containing mistaken answers as long as the right questions have been addressed, and not regard such issues as being an excess of jurisdiction (*Sherwood & Casson* v *Mackenzie* [1999]). The courts (*KNS* v *Sindall* [2000]) have also decided that where an Adjudicator allows a deduction in its decision, even though no effective notice of withholding has been given, the deduction would be an error, not an excess of jurisdiction.

The jurisdiction issue has arisen in many of the appeals to the courts against adjudication decisions made by 'losing' Parties (see many of the Selected Adjudication Summaries). Jurisdiction cases may be grouped under the following three main headings.

Jurisdiction given in adjudication clauses

The first situation is when the adjudication procedures expressly or implicitly permit that an Adjudicator can decide certain matters of jurisdiction. Contractual adjudication procedures can certainly do so, as long as the eight minimum requirements of the Construction Act are not diminished for what are intended to be Construction Act-compliant procedures. Even the Scheme, in limited circumstances, can let an Adjudicator decide jurisdictional matters. For example, it expressly permits an Adjudicator to decide whether two disputes are substantially the same (and if so the Adjudicator must resign). The courts (*Sherwood & Casson* v *Mackenzie* [1999]) agreed that was correct and supported an Adjudicator's decision that two disputes were not substantially the same, stating that they required substantial grounds before supporting any alternative view that the Adjudicator was wrong.

A different issue is whether an Adjudicator can decide the terms of a construction contract, e.g. the interpretation of non-jurisdictional conditions of contract, or measurement or payment provisions. The courts (*Butler* v *Merewood Homes* [2000]) stated that the Scheme included that an Adjudicator had jurisdiction to decide such matters and, whilst acknowledging that the Adjudicator may have decided wrongly, still supported that decision. In a similar case (*LPL* v *Kershaw* [2001]), the courts agreed that to interpret the meaning of a contract was within an Adjudicator's jurisdiction, and that they would support the decision even if it was an incorrect interpretation.

Parties decide on/agree to Adjudicator's jurisdiction

The second situation is where there is no agreement within adjudication procedures for an Adjudicator to determine matters of jurisdiction, but the Parties decide, or agree between themselves, that is what they want. For example, in one case (*Northern Developments* v *Nichol* [2000]), two Parties agreed to a Scheme amendment whereby the Adjudicator could award costs against the losing Party (whichever of the Parties this turned out to be). There was never any suggestion made that this was not agreed by either Party or that the Adjudicator did not have jurisdiction to make such a decision. It was therefore supported by the courts.

In another case (*Nolan Davis* v *Catton* [2000]), the Parties expressly asked an Adjudicator to decide a dispute which included determining who the Parties to a construction contract actually were. The Adjudicator's decision was supported by the courts as there was a joint agreement to accept the decision, even though in effect this included the Adjudicator deciding its own jurisdiction.

In a judicial review (*Watson* [2001]), the courts agreed that Parties could appoint an Adjudicator specifically to determine the correct interpretation of sub-contract terms that impinged directly on its jurisdiction and appointment.

Adjudicator decides its own jurisdiction

The third situation is where there is no agreement whatsoever to let an Adjudicator decide a jurisdictional issue, but an Adjudicator does so nevertheless. In this situation, it is more likely that the decision will be challenged by the losing Party in court. The courts will decide the issue on the merits of the case (i.e. they may or may not support the Adjudicator). For example, an Adjudicator's decision that pipework was not part of 'plant' and that there was a resulting construction contract (and hence the Adjudicator had jurisdiction) was overturned by the courts (*Homer Burgess* v *Chirex* [1999]). The courts in this Scottish case further stated that it was up to them whether they separated and enforced what they nevertheless considered were valid parts of a decision. This is at variance with the English courts (*Farebrother* v *Frogmore* [2001]) but supported by a later Scottish case (*Barr* v *Law Mining* [2001]).

In another case, a disagreement over the interpretation of the terms of a settlement was returned to the original Adjudicator, who decided that the settlement itself was a construction contract and hence the Adjudicator had jurisdiction. This was appealed and the courts decided it was too complex for a summary judgment and the issue had to go to trial to be sorted out (*Lathom* v *Cross* [1999]).

A further dispute (*Grovedeck* v *Capital Demolition* [2000]) concerned the existence or otherwise of a construction contract under the Act at the time of referral to an Adjudicator. It was agreed there were oral agreements between the Parties, but it was always refuted by the Responding Party that these were intended to create a written contract under the Act. However, the Adjudicator decided there was a written contract and thus effectively decided on its own jurisdiction. On appeal, the courts agreed that there was no construction contract at the time of Adjudicator appointment and so no jurisdiction existed.

On an 'internal' jurisdiction issue (*Carter* v *Nuttall* [2000]) the courts (*Karl Construction* v *Sweeney* [2000] and on appeal [2002]) attempted to clarify the extent to which an Adjudicator can stray away from the referred dispute in reaching a decision on that dispute. In this case the courts agreed that the Adjudicator could address issues that had not been raised specifically in the referral, if reaching a decision on those issues was directly relevant to reaching a decision on the questions that were being referred. The courts considered that the Adjudicator should have advised the Parties that it was considering such additional contract interpretation issues so that they could make submissions to the Adjudicator on them. However, this was a procedural 'mistake' inside jurisdiction, which did not prevent the adjudication decision from being enforced. In this case the courts clearly believed the Adjudicator had not strayed too far from its natural justice obligations. The latter is likely to be a developing area of law.

Umbrella Group Draft Guidance to Adjudicators using the Scheme

Challenges to jurisdiction

The issue

The Umbrella Group identify the issue as follows:

> *Almost the only way that a party can stop an adjudication proceeding is by alleging that the Adjudicator lacks the jurisdiction – that is, the authority – to deal with the issues raised. If an Adjudicator proceeds without the necessary jurisdiction his decision will not be enforceable. On the other hand, if he fails to proceed when he does have jurisdiction, that is unjust to the referring Party.*

Examples of grounds on which jurisdictional challenges may be made are:

- *the contract is not a 'construction contract' within the Act (s. 105 & 107)*
- *the relevant activities are not 'construction operations' within the Act (s. 105)*
- *the Adjudicator was not properly appointed, for example, because he is not the Adjudicator named in the contract, the wrong ANB made the appointment, the appointment was too late, or the Adjudicator has a conflict of interests*
- *it is asserted that there is no dispute.*

The law

The Umbrella Group describe the legal position as follows:

> *Unless the parties agree otherwise, an Adjudicator does not have the power to make a final decision as to whether he has jurisdiction to act as an Adjudicator under the Scheme; only the courts can do that. However, as one judge put it (*Christiani & Neilsen Ltd *v* The Lowry Centre Development Company Ltd*): 'It is clearly prudent, indeed desirable, for an Adjudicator faced with a jurisdictional challenge which is not a frivolous one to investigate his own jurisdiction and to reach his own non-binding conclusion as to that challenge'. An Adjudicator would find it hard to comply with the statutory duty of impartiality if he or she ignored such a challenge.*

The guidance

The Umbrella Group suggest the following:

1. If an Adjudicator is faced with a jurisdictional challenge, then the Adjudicator should investigate, seek the views of the Parties and reach its own conclusion on the merits of the challenge. If an Adjudicator fails to do so then it may seem that the Adjudicator is not impartial.
2. If the Adjudicator concludes that jurisdiction does exist, then the Adjudicator should tell the Parties immediately and continue with the adjudication.
3. If the Adjudicator concludes that jurisdiction does not exist, then the Adjudicator should tell the Parties immediately and give notice in writing of the Adjudicator's intention to resign. (Paragraph 9 of the Scheme.)
4. If the Adjudicator is unsure whether or not jurisdiction exists, the Adjudicator should nevertheless make a judgement and a decision as to whether to proceed or whether to resign.
5. If the Adjudicator does proceed, the Adjudicator should [presumably to ensure that the Adjudicator's fees will be paid] consider obtaining confirmation from the Referring Party that it wishes the Adjudicator to continue.

Summary

In summary, where an Adjudicator's jurisdiction is agreed by the Parties, including by it being incorporated expressly or implicitly within the adjudication provisions of the construction contract, that jurisdiction will be supported by the courts as well as any resulting decision, even if wrong (*Bouygues* v *Dahl-Jensen* [1999]). However, if the Adjudicator is thought by a Party to have acted outside its agreed terms of reference that is ultimately a matter to be resolved by the courts, not the Adjudicator.

Where there is no agreement regarding Adjudicator jurisdiction, it must be anticipated that any Adjudicator's decision is likely to be appealed by the losing Party and the courts will be asked to decide the jurisdiction issue. It is suggested that 'threshold' jurisdictional issues, i.e. those that go to the core of whether there should be adjudication or not (*Carter* v *Nuttall* [2000]) should be undertaken as soon as possible, e.g. before the adjudication process is significantly advanced. The courts will support a rapid resolution of such matters (*Palmers* v *ABB Power* [1999]). Whatever else, a Party cannot expect to be able to raise a 'threshold' issue as a defence at enforcement proceedings if it hasn't raised (and probably maintained) it before, or in a timely manner during, the adjudication itself (*Maymac* v *Faraday* [2000]).

It is interesting to contemplate the implications of an Adjudicator being asked by the Parties to decide a dispute on whether another Adjudicator had jurisdiction in a different dispute. Presumably the Parties could give such an instruction and incorporate a statement that they agreed to be bound by that Adjudicator's decision on jurisdiction. As a final and binding agreement between the Parties, would the courts interfere even if they thought that Adjudicator's decision was by law incorrect? It is probable that the process would be considered to be one arising 'in connection with' rather than 'under' a contract and would therefore not be governed by the requirements of the Construction Act. Could the Parties therefore agree that the Adjudicator could use any adjudication procedures they wanted it to use to determine this jurisdictional issue? Currently the JCT has set up a panel of Arbitrators to decide jurisdictional issues. Why shouldn't that role be one for an adjudication panel or even a single appointed Adjudicator to sort out?

Chapter 11

Procedural fairness

Introduction

The Construction Act requires an Adjudicator to 'act impartially' but there is no express (i.e. written) requirement in the Construction Act for an Adjudicator, unlike an Arbitrator, to be 'fair' or 'act fairly'. (To be pedantic, it will be noted that an Adjudicator could, in theory, be impartially fair or impartially unfair to both Parties, i.e. be impartial whether or not it is fair.) However, few would argue that Adjudicators should, in both Construction Act-compliant and non-compliant adjudications, proceed in a manner fair to the Parties, as far as they are able in any particular set of circumstances. It is up to the Adjudicator to determine in each case the most appropriate way to do so in order to arrive at a proper decision within the time stipulated. An Adjudicator (perhaps assisted by specialist advice where necessary) must recognise what is appropriate in any given situation.

What constitutes procedural fairness is not neatly encapsulated in either a contract or a statute. Instead, it comes from the application in practice of sets of rules.

One set of rules is the European Convention on Human Rights, which is incorporated into the Human Rights Act 1998. The latter is discussed later within this chapter.

Another set of rules is what is known as the Rules of Natural Justice. Most developed countries have developed such sets of rules and England is no exception. These rules are contained in a body of decided cases under the common law. Contractual adjudications, and statutory Scheme adjudications, (although there may be additional requirements (e.g. see Paragraphs 4, 10 and 17 of the Scheme)), must still comply with common law.

For example, one common law requirement is that each Party should know what the other Party is saying and be given an opportunity to respond. As a consequence, it is usual for Adjudicators to issue instructions that every letter sent to the Adjudicator by a Party must be copied to the other Party and that every contact with the Adjudicator should be recorded and copied to both Parties. Thereafter, where an Adjudicator reaches its decision based solely on documentation and has no other contact with either Party, there should be no problem. Similarly, where there is little documentation but the decision is based on a meeting or meetings held with all Parties present to make their own record of what is discussed, the position should also be relatively simple. With regard to this one requirement for fairness, it is situations in between the two extremes that can give rise to difficulties.

There are other procedures that need to be seen to be being tackled fairly by an Adjudicator. An example would be any factual issues that an Adjudicator discovers in the course of an adjudication. Where these may be relevant to the decision and one or other of the Parties are unaware of them, an Adjudicator must put them to the Party or Parties for their comments. Whatever the common law position is, it will be noted that the Scheme, Paragraph 17 states: 'The Adjudicator ... shall make available [to the Parties] ... any information to be taken into account in reaching the decision.' This thus imposes a specific statutory duty on the Adjudicator. This rule is so fundamental that the courts have suggested that even if not expressly incorporated into contractual adjudication procedures, it is likely to be implied into them by law. Be that as it may, in the case of statutory Scheme adjudications, the provision is an express one. It even means, for instance, that if an Adjudicator is aware of a relevant court case that the Adjudicator intends to rely upon in reaching its decision, then the Adjudicator is under an express statutory obligation to mention that case to the Parties so that they can respond if they want to.

Human Rights Act 1998

The Human Rights Act 1998 came into force on 2 October 2000. It incorporated the European Convention on Human Rights into English law.

Article 6 applies to decision-making bodies that are public authorities and to persons, 'certain of whose functions are of a public nature', i.e. it states how such authorities and persons must conduct themselves when dealing with others.

Article 6.1 states that 'In the determination of his civil rights and obligations ... everyone is entitled to a fair and public hearing within a reasonable time by an independent and impartial tribunal established by law'.

A tribunal should therefore hold its proceedings in public and its decision must be made public. European case law is already well established and is applicable. This includes the right to the disclosure of documents, for reasons to be given with decisions and 'equality of arms', which has been explained as 'each Party must be offered a reasonable opportunity to present its case, including his evidence, under conditions that do not place him at a substantive disadvantage *vis-à-vis* his opponent'.

How does this apply to adjudication?

A public body includes anyone fulfilling a public function. Adjudication under the Construction Act is a statutory process that is compulsory for a Responding Party once a Referring Party has selected the option to have its dispute adjudicated in accordance with a procedure complying with the Act. It has therefore been claimed that Adjudicators and Adjudicator nominating organisations are public bodies. This presumably would not apply where adjudication was outside the statutory process, e.g. there was no 'construction operation' in accordance with the Act.

Certain English cases have clarified the situation.

The courts (*Elenay* v *The Vestry* [2000]), stated that '... Article 6 does not apply to an Adjudicator's award or to proceedings before an Adjudicator, because although they are a decision or determination of a question of civil rights, they are not in any sense a final determination'. The courts disagreed that human rights might have been infringed because the limited timetable for an adjudication to produce a decision had not allowed one Party a reasonable time to present its case and this was not fair.

Two issues arise from this:

(a) It has been suggested that there is an obvious case of a breach of civil rights if, as a consequence of following an Adjudicator's decision, a Party in the interim, before having its dispute heard in litigation or arbitration, becomes bankrupt or incurs similar serious consequences, without having an opportunity to fully present its case.
(b) There will be situations where the Parties pre- or post-decision agree that decisions will be final and binding, or there are contractual clauses that make it so if neither Party follows certain subsequent procedures within a certain timescale. It is suggested that if the Parties expressly or implicitly accept a decision as final and binding they must be free to do so, without later being able to use that as an excuse to appeal on the basis of a breach of the Human Rights Act.

The courts (*Austin Hall* v *Buckland* [2001]) have also stated that:

- An Adjudicator on a Construction Act adjudication is not a public authority and so is not bound by the Human Rights Act.
- An Adjudicator is not a person before whom legal proceedings could be brought, according to the definition of tribunal given in Section 21 of the Human Rights Act.
- Although under the Human Rights Act everyone is entitled to a public hearing, if a Party involved in adjudication does not ask for publicity, it cannot then claim non-compliance with the Human Rights Act on that basis after it loses. (In other words, when a Party agrees to waive an issue that would otherwise be a non-compliance with a requirement of the Human Rights Act, it appears that later it cannot change its mind. It will be noted that this is similar to the Parties agreeing that a decision will be final and binding.)

In summary, therefore, it is considered that the European Convention of Human Rights does not apply to adjudications. That position could change in the future.

Rules of Natural Justice

There is no express requirement in Section 108 (2)(e) of the Construction Act for 'natural justice'. This does not mean, however, that there is no requirement for an Adjudicator to observe the requirements for natural justice. Indeed adjudication procedures (e.g. the Scheme Paragraphs 12 and 17) might not expressly state the words 'natural justice' but can nevertheless mean something very similar.

It is useful to discuss what the Rules of Natural Justice in England actually may be. Essentially they require that:

1. No one should judge him or herself.
   For adjudication, this will usually relate to an Adjudicator deciding its own jurisdiction (see Chapter 10). A Responding Party will always be looking for a way to object to jurisdiction as a potential first line of appeal against an adverse adjudication decision. Such objections will sometimes appear to an Adjudicator to be, and may actually be, on somewhat spurious grounds. For simple undefined objections, an Adjudicator will often be able to assess whether the adjudication and associated appointment have been made, for instance, in accordance with the contract, that there is a 'construction contract' (where applicable) and if authority exists to deal with a particular dispute. Most objections will be more specific regarding why jurisdiction is being objected to. An Adjudicator will often then be able to check the validity of those objections. If satisfied there is no obvious problem, and usually having the support of the Referring Party for such a procedure, an Adjudicator can acknowledge receipt of any objection and nevertheless proceed to meet the adjudication timetable. If there is only one objecting Party, the latter can register its objection and let the adjudication proceed, with that Party's involvement being stated clearly as being on a without prejudice basis, leaving the issue to be sorted out later (*Project Consultancy* v *Gray Trust* [1999]). Otherwise, if an adjudication came to a halt every time there was even a hint or speculative suggestion of lack of jurisdiction by one of the Parties, because the Adjudicator cannot decide its own jurisdiction, this would be a successful ploy to stop every adjudication with a reluctant Party. This would clearly be completely counter to the Construction Act. The option is always available for the Parties to approach the courts for a quick decision on jurisdiction before further time and money, possibly including costs (*Atlas Ceiling* v *Crowngate* [2000]), is spent (*Palmers* v *ABB Power* [1999]).
2. Each party must have a reasonable opportunity of putting its case and dealing with that of its opponent.
   Problems can arise in adjudication due to the limited timescale, e.g.:
   - A Responding Party has a limited time in which to respond to what might be a detailed complex submission by a Referring Party. Further, or alternatively, a Responding Party provides a complex response that the Referring Party wishes to consider and probably reply to but has no time available to do so. This would not appear to be reasonable.
   - A Responding Party might submit a significant directly relevant counter-claim (or set-off or abatement) defence which may comply with the payment notice, provisions of Sections 110 and 111 of the Construction Act. With a limited timescale (even if extended), the same problem of reasonableness applies, and additionally an Adjudicator may not have time to consider accepting any response to the counter-claim if the latter was provided. If an Adjudicator agrees to deal with the counter-claim in such circumstances, this again would appear to be contrary to natural justice. Conversely, if the Adjudicator refused to deal with the counter-claim, particularly one complying

with notice provisions of Sections 110 and 111, it could again be accused of acting contrary to natural justice.

- There might be later submissions or amplifications, provided by either Party, that similarly cannot be addressed, or addressed adequately, without extensions of time that might not be agreeable to the Adjudicator or one of the Parties.

Timetable restrictions afford many Adjudicators with problems, particularly in their attempts to be fair. A robust approach to the acceptance of later submissions, including requiring extensions of time in some circumstances, can assist. However, for Scheme adjudications, many Adjudicators interpret Paragraph 17 as meaning that they have to accept, and in theory consider, submissions received right up until their decisions are signed.

3. Matters which could influence a decision, insofar as they are specific matters, should be made available, and in reasonable time, for the comments and submissions of the Parties. (See also 'Receipt of without prejudice confidential information' in Chapter 12.) A decision should not be based on specific matters which either or both of the Parties have never had the chance to deal with, nor is it right that a Party should learn first of adverse points in the decision against it. For example:
   - If an Adjudicator has found significance in a point which has never been raised by either side, then the Adjudicator must put it to the Parties so that they have an opportunity to comment.
   - An Adjudicator cannot consider statements or submissions provided by one Party which have not been made available to the other Party.
   - The Parties should be provided with the results of any relevant investigations/information obtained, by the Adjudicator.
   - The Parties should be advised of any analysis of facts intended by the Adjudicator if this is alternative to, or additional to, anything previously undertaken by the Parties in connection with the referral, and the results of that analysis (*Balfour Beatty* v *Lambeth* [2002]).

There would seem to be nothing unreasonable about the foregoing. It is suggested that, time permitting, an Adjudicator should always comply, whether or not there is an express requirement to follow the Rules of Natural Justice or stated requirements which amount to something similar.

In at least one directly relevant Scheme case, the courts (*Woods Hardwick* v *Chiltern* [2000]) decided that the Adjudicator had not kept one Party properly informed of the results of evidence obtained from others. Because of this (and other reasons) the courts would not support a request to enforce that Adjudicator's decision.

In another case, the courts (*Discain* v *Opecprime* [2001]) stated that they would not enforce an Adjudicator's decision where the Adjudicator had substantially breached the Rules of Natural Justice. The breach was a failure by the Adjudicator to inform one Party of significant submissions made to it by the other Party. In fact, it appears that the Adjudicator was relying on the first Party to notify the other Party of what had taken place and in one instance there was delay of 3 days in doing so. (The conclusion must be that to ensure minimum delay, the Adjudicator must take responsibility for informing the other Party). Whilst appreciating the difficulties associated with the limited timetable, the courts decided the Adjudicator had 'overstretched' the Rules of Natural Justice. They identified two types of breach of natural justice, i.e. those that had no discernible effect on the decision and those that might have a significant effect (as in this case). It seems probable that if the fax had been sent to the second Party on the same day as the communication between the Adjudicator and the first Party, the courts might have decided there was no breach. The only real safeguard for the Adjudicator in such circumstances would be for it to send out details of submissions and discussions itself, and as soon as reasonably possible.

Partiality

Most contractual adjudication procedures do not specifically state that an Adjudicator must be impartial, only to act impartially, but it is not clear whether the courts would differentiate between the two. As the Construction Act does not specifically state that an Adjudicator should be independent, it does not, in theory, preclude the selection of the contract Architect or Engineer to act as an Adjudicator.

A court's test for partiality in arbitration (*R* v *Gough* [1993]) is whether an Arbitrator 'might unfairly regard or have unfairly regarded with favour or disfavour the case of a party', considering all the evidence known to the court at the time the question is considered. The word 'might' in the quotation refers to 'possibility rather than probability of bias'. It is considered that this would equally apply to adjudication, and would almost certainly preclude construction professionals engaged by a Party to a dispute, such as an Employer.

The appearance of impartiality must be maintained, even in the face of a belligerent Party who might be abusive, confrontational and inexperienced according to the Adjudicator (*Woods Hardwick* v *Chiltern* [2000]), or who might be deliberately obstructive to the adjudication process by being slow or more reactive to Adjudicator requests, or in complying with the adjudication process. In the adjudication case the Adjudicator, at the request of Woods Hardwick, produced a witness statement to assist Woods Hardwick in the enforcement procedure, which contained the Adjudicator's impressions of Chiltern. Whilst it was not illegal to produce a witness statement, the courts decided that in this case, the witness statement suggested a lack of impartiality might have existed. For this, and other reasons, the courts refused to enforce the Adjudicator's decision.

Bias

The Court of Appeal (*Locabail (UK)* v *Bayfield* [2000]) is the leading judgment on bias. It reviewed the three types of bias applicable to tribunals, i.e. actual bias, having an interest in the outcome of a case, and unconscious bias.

On actual bias, the Court said 'All legal arbiters are bound to apply the law as they understand it to the facts of the individual cases as they find them. They must do so without fear or favours, affection or ill-will, that is without partiality or prejudice.'

On having an interest in the outcome of a case, the Court said this bias arose when a judge 'is a party to the case or has a pecuniary or proprietary interest in it or ... is so closely connected with a party to the proceedings that he may be said to be acting in his own cause.'

On unconscious bias, the courts have been traditionally concerned with a legitimate fear of bias undermining confidence in judicial procedures. The test used by the courts for unconscious bias is whether there is a 'real danger of bias' and this has been stated by the courts to apply equally to arbitration, even though the Arbitration Act 1996 refers to 'justifiable doubts'.

An appearance and claim of bias may result from an Adjudicator having a continuing or frequent social, or probably more important, contractual relationship with a Party.

Paragraph 4 of the Scheme states that the Adjudicator shall actually *declare* any interest, financial or otherwise, in any matter relating to the dispute. Thus, an Engineer engaged by an Employer or sometimes bidding for work with that Employer, would presumably have a declarable interest. It will also be noted that Paragraph 10 of the Scheme states that an objection to an Adjudicator's appointment shall not invalidate the appointment or the decision. Therefore, whether an interest has been declared or not (or one was later found out by a Party that considered that it should have been declared), the most a Party might do is object, in which case Paragraph 10 applies.

The foregoing only relates to the statutory Scheme, of course. However, it is considered that, unless particular adjudication contractual procedures give express alternative requirements or clarifications, the safest procedure to avoid objection or invalidation would appear to be to identify and declare any possible perception of a conflict of interest as soon as possible. At the very least, if a potential interest is declared, and there is no early response from either Party, the courts would be less likely to overrule a decision if an objection was only raised at the time of that decision.

The courts (*Glencot* v *Barrett* [2001]), which also gave a detailed review of the law of bias, stated that where an Adjudicator has also acted in a mediation role between the same Parties, there might be perceived bias. If this was claimed by a Party, it would thus need to be considered by the courts in any appeal on the enforcement of an Adjudicator's decision. This case is discussed further under the heading 'Interim mediation by an Adjudicator' in Chapter 14 Miscellaneous Issues.

Another case (*Mowlem* v *Hydra-Tight* [2000]) essentially concerned a dispute on the validity of a contractual requirement to select an Adjudicator from a pre-agreed list of Adjudicators provided within the conditions of the contract. The courts supported the validity of this approach on the basis that the contract was freely entered into by both Parties. This would not, of course, prevent a Party from objecting to an Adjudicator chosen in this way, if there was perceived bias or partiality during the adjudication process itself.

A further case (*Balfour Beatty* v *Lambeth* [2002])concerned a situation where an Adjudicator, in effect, made good by additional analysis (and without discussion with the Responding Party), the deficiencies in a Referring Party's case.

Another situation that might arise is where an Adjudicator and a professional advising one of the Parties belong to the same organisation or even work in the same set of offices. How effective are 'Chinese walls'? The courts in an arbitration (*Laker Airways* v *FLS Aerospace*) found that such a conflict of interest only actually arose where the same person undertook conflicting duties for different clients or put itself in a position where a duty to a client conflicted with its own self interest.

Independence

The latter would appear to be exactly the situation that arises where an Engineer or Architect acts, in effect, as an Employer's representative (possibly providing employees to administer a contract on an Employer's behalf) and the same person is then supposed to provide impartial decisions on disputes arising between the same Employer and its Contractors. Those disputes might even be the alleged consequence of actions, errors or omissions by the Engineer or architect whilst administering the contract. This perceived lack of independence was one of the reasons for introducing statutory adjudication.

Article 6.1 of the European Convention on Human Rights expressly requires that any tribunal determining civil rights be independent. It is thus considered that independence of an Adjudicator is a basic requirement. It will be noted that some contractual adjudication procedures actually expressly require that an Adjudicator be independent (e.g. GC/Works Procedure, Paragraph 6).

There is, however, no provision for independence in either the Construction Act or the Scheme. When being drafted, it was suggested to the Government that independence be included as a requirement but impartiality (a concept different from independence) was as far as the drafters were prepared to go. In fact, as drafted, the Scheme does not even require the Adjudicator to be impartial, simply to *act* impartially. What that means in practice is that an Adjudicator regularly retained as, say, a consultant by a Referring Party (and, in the extreme, whose entire income may depend upon the latter's goodwill) could nevertheless be an Adjudicator provided it acts impartially.

In summary, an Adjudicator is obliged, at least for statutory adjudications, to act impartially and, it is suggested, must comply, as far as reasonably possible within the confines of the pertaining timetable and relevant procedures, with the Rules of Natural Justice and be fair. If either or both natural justice and to be fair are express requirements of contractual adjudication procedures or conditions of contract, they are then, of course, contractual obligations.

Umbrella Group Draft Guidance to Adjudicators using the Scheme

Human Rights

No guidance is given.

Natural Justice (Procedural fairness)

The issue

The Umbrella Group identify the issue in this area as being as follows:

> *An Adjudicator must conduct the proceedings in accordance with the requirements of natural justice or procedural fairness. In a small number*

of cases, the courts have not enforced the Adjudicator's decision on the grounds that the Adjudicator did not act fairly, and it is evident that Adjudicators do not always know exactly what it means to act in accordance with natural justice.

The law

The Umbrella Group identify the law in England in this area to be as follows:

> *Natural justice is not a defined term. As one judge (Lord Reid in* Wiseman *v* Borneman *[1971] AC 297 HL) put it: 'Natural justice requires that the procedure before any tribunal which is acting judicially shall be fair in all the circumstances, and I would be very sorry to see this fundamental principle degenerated into hard-and-fast rules.'*

There are two main aspects to the need for procedural fairness: 'no bias' and 'fair hearing'.

No bias: the Adjudicator must be impartial and act independently. For there to be a breach of natural justice it is not necessary for there to be actual bias – apparent bias is sufficient. The test is whether there is a real possiblity, *not probability, that the Adjudicator is biased based on how the reasonable observer would interpret the situation. Bias may occur in a number of ways; for example, if the Adjudicator*

- *has, or appears to have, a personal relationship with one of the parties*
- *has, or appears to have, an interest in the outcome of the adjudication (a conflict of interests)*
- *conducts the adjudication in a manner which favours, or seems to favour, one party*
- *acts in a manner which is seen, or might be seen, as supporting one party to the detriment of the other.*

Fair hearing: this means ensuring that each party:

- *has a reasonable opportunity of presenting its case;*
- *knows what the case is against it;*
- *is in possession of all the evidence and information that is adduced against it or obtained by the Adjudicator.*

The words do not imply that an oral hearing or meeting is necessary. These requirements have to be measured in the context of the time within which the Adjudicator has to reach its decision, and the fact that the decision is provisional, pending final resolution of the dispute by arbitration, litigation or agreement. As one judge (Judge Humphrey Lloyd in Glencot *v* Barrett *[2001]) put it: 'It is accepted that the Adjudicator has to conduct the proceedings in accordance with the Rules of Natural Justice or as fairly as the limitations imposed by Parliament will permit.' (See also Judge Bowsher in* Discain *v* Opecprime *[2001].)*

The guidance

The Umbrella Group suggests the following:

1. If Adjudicators are aware of any connection, however remote, that they have or have had with either Party, or of any matter which either Party might see as being a conflict of interests, then they should notify the Parties and consider refusing the appointment. If anything comes to light after the Adjudicator has been appointed, the Adjudicator should ask the Parties if they wish to continue and, if necessary, the Adjudicator should resign. (Paragraph 9(i) of Scheme.)
2. Adjudicators should ensure that all their actions are, and are seen to be, fair. Before taking any action (for example, declining to accept late information (Paragraph 15 of Scheme)) Adjudicators should ask themselves: 'Am I acting fairly? Does it appear that I am acting unfairly?'
3. Although the Scheme does not make express provision for the submission of a response to the Referral Notice, Adjudicators should give the Responding Party a reasonable opportunity to respond, and consider any response that is served. (Paragraph 17 of Scheme.) Adjudicators may also require that a response is served, and specify any timetable for this to be done. (Paragraph 13(a) of Scheme.)
4. Adjudicators should use great care if communicating with one Party in the absence of the other either in a meeting or over the telephone. If contact with one Party alone is necessary or unavoidable, Adjudicators should keep a detailed record of what is said and send it to both Parties as soon as practicable. If Adjudicators receive unsolicited telephone calls, they should consider requiring that the information is conveyed in a form (for example, fax or e-mail) which can be sent to the other Party and the Adjudicator. Adjudicators should ensure that both Parties know what is happening at all times.
5. In the case of necessary telephone calls, Adjudicators should consider using telephone conferencing.
6. Adjudicators should consider obtaining the views of the Parties before making directions other than those which the Adjudicator considers routine.
7. Adjudicators should notify deadlines for the supply of information and dates for meetings as soon as possible.
8. Although their role is investigatory and inquisitorial, when undertaking investigations, Adjudicators should ask themselves whether they are attempting to make or supplement one Party's case. Adjudicators should not argue the case for one of the Parties, either before or when giving the decision, or in setting out reasons.
9. Adjudicators should consider carefully before indicating disapproval of the way in which one Party conducts its case.
10. Adjudicators should guard against seeming to have made up their minds as to the merits of the case before having heard or seen all the evidence.
11. Adjudicators should ensure that any information on which they intend to rely in reaching their decision is made known to both Parties, so that both Parties have an opportunity of responding.

Chapter 12

Conduct of the adjudication

Introduction

Although the Construction Act creates a right to adjudication where it did not exist before, it is the underlying construction contract and adjudication procedures (whether incorporated directly or by reference) that provide the Adjudicator with its powers and create the rules that must be followed.

There is now a significant body of case law relating to and interpreting the procedural requirements of the Construction Act and the statutory Scheme of which an Adjudicator must be aware.

There has currently been much less case law relating to contractual adjudications, other than those relating to natural justice, and an Adjudicator will be far more guided by procedures stated in the adjudication procedures. Thus, as long as contractual adjudications comply with natural justice (as far as they reasonably can), the Parties are free to decide what disputes they want adjudicated and the procedures they want adopted. In these circumstances, an Adjudicator will be bound by the terms of the contract and case law is less likely to be relevant.

This chapter deals with typical procedural issues that might arise and the typical provisions contained within adjudication procedures or within relevant adjudication clauses of conditions of contract.

Particular adjudication procedures should be referred to, where necessary, for specific procedures and wordings. For example, whereas Section 108 (2)(f) of the Construction Act states that a contract should 'enable' an Adjudicator to take the initiative in ascertaining the facts and law required to make a decision, some adjudication procedures use words such as 'shall' or 'may'. If it is 'shall' take the initiative, it is mandatory, notwithstanding any difficulties that might arise due to the limited time usually available. If a word such as 'may' is used, an Adjudicator can decide what is appropriate and reasonable in the available timescale. For instance, adjudication procedures will normally not preclude (or alternatively might insist on) an Adjudicator being adversarial or inquisitorial, being Proactive, following a line of enquiry with the Parties (even with third parties), deciding whether there is a need for external legal or technical advice, a meeting, a site visit and so on.

Adjudication timetable

In practice, even for contractual adjudication, there will usually be a 28-day adjudication timetable, and any formal extensions to it, that will probably have the maximum overall impact on procedures adopted by an Adjudicator. It will be seen in the text, including this chapter, that available time will be a key issue in instructions given and procedural decisions made.

Initial planning

After receiving the referral documents, an Adjudicator should be able to consider a number of issues, including:

- How long the Responding Party can have to reply, if not specified in the adjudication procedures.
- Whether to have a meeting and with whom.
- Whether a site visit is required, whether accompanied by representatives of one or both of the Parties, and if so by which particular representatives.
- Whether external advice is necessary.
- Whether some new, additional or alternative analysis of facts is intended by the Adjudicator that might need discussion with the Parties (see *Balfour Beatty* v *Lambeth* [2002]).
- Whether in-house assistance to the Adjudicator (where available) might be beneficial.
- How long it is realistically likely to take to reach a decision.
- If a possible extension of time for a decision should be discussed.

The Adjudicator should then write to the Parties stating the adjudication timetable and other requirements.

An example Adjudicator's first letter is given in Appendix 2.

Appendix 3, provided in this text as an attachment to the first letter (although it could form an integral part of the first, or be a separate, letter), lists some possible first Adjudicator Instructions.

Sometimes Parties may be unsophisticated in adjudication matters, particularly where they are individuals or small organisations with no regular access to experienced outside advisors, or one Party might be significantly more sophisticated than the other. It is suggested that in such circumstances it would be in both the Parties' and Adjudicator's interests for the latter to provide some written advice on how the Adjudicator intended to proceed with the adjudication, the stated powers of the Adjudicator and so on. Alternatively, or to support this assistance, the Parties could be referred to advisory notes sometimes issued by bodies involved in adjudication procedures. Also, no disadvantage is perceived to the Parties or the Adjudicator if an Adjudicator points out some specific relevant contract or statutory clauses. An example is Clause 66 (9)(b) of the ICE Conditions of Contract, which states that where an Adjudicator has given a decision in respect of a particular dispute, any notice to refer to arbitration must be served within three months of the giving of the adjudication decision, otherwise the latter would be final and binding. Appendix 4, also written as if it is an attachment to the Adjudicator's first letter (although again it could form an integral part of the first, or be a separate, letter), lists some possible examples of advice that could be provided.

Limiting the extent of the referral documents

There is no express statement in the Construction Act which limits referral documents, only a requirement for Adjudicator impartiality (i.e. to treat both Parties in a similar way as far as is reasonably possible, within any confines such as time limits set by the adjudication). There is therefore no reason why the underlying construction contract or adjudication procedures cannot limit the amount of documentation submitted to an Adjudicator. For example, the Scheme, Paragraph 13(g), with the support of Paragraphs 14 and 15, permits an Adjudicator to instruct a limit to the length of written documents (although there is no guidance on the limits). There is usually no precise equivalent option for the Adjudicator in contractual adjudication procedures, but neither are there express exclusions. For example, the ICE Procedure does not mention a limit but an Adjudicator might give an instruction using Paragraph 5.5 powers (bearing in mind the obligation to be both impartial and additionally fair in the ICE Procedure).

There are obvious advantages to the Adjudicator in limiting the size/extent of the referral document bundle, and it also concentrates the Referring Party's mind and reduces the extent of possible 'ambushing' (see later in this chapter) of the Responding Party with a large number of documents to study and respond to in a limited time. A limit of 5,000 words was regularly referred to in pre-Scheme documents discussing adjudication, and this might be a useful guideline.

However, an Adjudicator has no control over an adjudication until after its appointment, which would probably be very close to when, or after, the referral documents had been received by the Adjudicator. This would be too late in practice to instruct the Referring Party to limit the size of its initial submission. If the latter has forwarded a large number of documents in support of its case, it would not be impartial as required by the Construction Act (and certainly not fair under contractual procedures such as the ICE Procedure) to impose any limitation on the later document bundle of the Responding Party.

One solution would be to include specific referral and responding document bundle limitations within the proposed adjudication procedures (although a view has been expressed that this would fall foul of the Human Rights Act 1998 (see Chapter 11) if it should be decided that adjudication does, after all, come under that Act). That would not appear to be a serious practical problem before the issue of tender documents, even if it involved the amendment of otherwise standard procedures. An alternative solution could be the insertion of a clause within the adjudication clauses of the conditions of contract themselves. For standard conditions of contract, it could be done by an appropriate 'Part II' type contract clause addition or amendment. For non-standard conditions of contract, a limiting clause could simply be included within the text. The same amount of care would need to be taken as would be necessary for any condition of contract clauses or amendments, but additionally it might be useful to ensure that the eight minimum requirements of Section 108 (2) to 108 (4) of the Construction Act (the obvious one being the requirement of the Construction Act for impartiality) were not breached to avoid providing a Referring Party with an option to use the statutory Scheme instead. Such an approach seems reasonable.

Whether to include a limit and, if so, what this should be, are basic questions that might prove more difficult to resolve. A partial, albeit practical, solution might be for an Adjudicator's first instruction to include a requirement, within a stated time limit, for the Referring Party to summarise its case on two or three A4 pages, cross-referring to the initial main referral documents. That same requirement could then be impartially (and fairly) imposed on the Responding Party.

'Ambushing'

In practice, a dispute usually arises when both Parties have previously presented their case to each other and one Party is dissatisfied with the resulting decision or lack of decisions within what it considers to be a reasonable time. In all disputes the Referring Party should therefore already have a full set of referral documents ready at the time of giving notice of its intention to refer a dispute to adjudication. The Responding Party should also have its documentation already at least partly prepared. A Responding Party will thus often be in a position to meet a specified time limit for its response, even though it may appear to have been suddenly presented with a large amount of Referring Party's information, i.e. it was 'ambushed'.

There will, however, be situations where a Referring Party presents an Adjudicator and Responding Party with a significant amount of what turns out to be new material, records and so on in support of its referral. A Responding Party might point out to an Adjudicator that this is so, that there is thus no dispute and so the Adjudicator has no jurisdiction. The courts (*Nuttall* v *Carter* [2002]) have agreed with this (see 'What is a dispute' in Chapter 14). At the very least the Party might ask that it should be allowed proper time to assess the new information. The point here is that once a Party has reviewed that new information it might change its mind regarding its earlier decision or delay in responding and, say, pay some or all of the money being claimed. In other words, there was not a dispute until the new material had been reviewed and a decision was reached that was still not acceptable to the Referring Party. With no dispute, the Adjudicator has no jurisdiction and hence cannot continue, at least on those parts of the dispute that the new material relates to.

This brings up some interesting issues. For example, on understanding the Adjudicator's problem, the Referring Party might ask if it could withdraw the new material and rely solely on the documentation already provided to the Responding Party. The Responding Party might not agree to this on the basis that the Adjudicator might subconsciously be affected by what it had seen (despite it being withdrawn) and this might affect its decision. It is suggested that if the latter occurs the Adjudicator does not agree to the Referring Party's request.

Another possibility is that both Parties agree an extension of time so that the Responding Party is given ample opportunity to review and respond to the new material. However, it will be realised that with no current dispute, the Adjudicator has no jurisdiction and strictly speaking is no longer involved in the process. Also, the end result might ultimately require a new notice of an intention to refer what has become a different dispute. Once again, the simplest solution is for the whole issue to be progressed outside what has become a very dubious adjudication and for the Adjudicator to cease being involved, other than perhaps through a later reappointment.

It is equally likely to be a Responding Party that provides the ambush material that the Referring Party says it has never seen before. It is suggested that the foregoing two issues will similarly apply in this situation.

There is another type of 'ambushing'. This is when several claimants (or one claimant with several separate disputes) deliberately arrange to serve their referral notices at the same time. The obvious option for a potential Adjudicator is not to accept more disputes (or enjoinders) than it can handle, after enquiring into available time extensions and considering how problems might be simplified by requesting summaries, as previously discussed.

Timetable for information submission

Due to the requirement to meet its own overall timetable for providing a decision, an Adjudicator should specify dates by which information from the Parties should be supplied or any other actions carried out. In many simple cases this will be the Responding Party, as the Referring Party should, at least in theory, have provided a full statement of its case in the referral documents. However, in many cases Adjudicators are finding that the Referring Party will want to reply to the Response, and then the Responding Party to that reply and so on until the decision is made. A robust Adjudicator might well draw a line at considering anything received after a certain date, taking into account its remaining availability in the days left before its decision. As long as that is done as fairly as possible within the time constraints imposed by the procedures, it should be supported by the courts in the case of an appeal. However, many Adjudicators, particularly those involved in Scheme adjudications, perhaps in trying to comply with the more onerous interpretation of Paragraph 17 (see Chapter 7 The Scheme) will continue to accept submissions up until the end. Adjudicators can do this, of course, if they have the time, although often they make the proviso that after a certain date they cannot guarantee they will have the time to actually deal with such late information (so much for Paragraph 17!).

Frivolous disputes or a non-cooperative Party

Contractual adjudication procedures are stated, or incorporated by reference, within conditions of contract, or the Scheme may become available in circumstances described elsewhere in this text and be selected by the Referring Party. Once an adjudication appointment is accepted, unless other matters such as termination intervene, an Adjudicator must proceed in accordance with those adjudication requirements to reach a decision. Occasionally, however, the actions or inactions of one of the Parties might make it difficult to achieve this.

An Adjudicator's first indications of a problem might be to receive no initial response from a Party at all, when one is expected. In that case, the first thing to do is to check that contact details are correct, that documents are being received and to try to establish communications. Alternatively, the first indications might be a letter from the other Party stating that it wants nothing to do with the adjudication. The Adjudicator must acknowledge receipt of the letter, advise that Party that the adjudication will be proceeding regardless and then make sure that Party is copied all documentation between the Adjudicator and the Referring Party, using a system where receipt of communications is recorded.

Sometimes a dispute might appear to be petty, one side might clearly have no defence, or a Party might acknowledge a liability but refuse to do something about it, perhaps as the consequence of a clash of personalities, but also perhaps a lack of finance or resources. Where there are several heads of claim under one dispute, the foregoing may apply to one or more of the heads of claim but not all of them. A problem is often that disproportionate amounts of time may have to be spent on the more minor frivolous issues.

The only scope for censure in most standard adjudication procedures is to award a larger proportion, or the whole, of the Adjudicator's fees and expenses (and costs in some adjudication procedures) against an erring Party, particularly if the final decision, or decisions, are against that Party. For example, some standard contractual adjudication procedures such as the ICE, CIC and JCT require the Parties to pay their own costs, but permit an Adjudicator in its decision to apportion its fees and expenses between the Parties as it sees fit. Other standard procedures such as GC/Works and the original Scheme do or may permit, and non-standard procedures can permit, an Adjudicator to award costs (see Chapter 13). A decision should be reached based on the applicable facts and law, although some procedures such as GC/Works/1 permit decisions (which would include decisions on fees and costs) on the basis of other matters, GC/Works/1 expressly referring to 'punishing unmeritorious' conduct in the Commentary on Condition 59. Otherwise, there is generally nothing expressly stated about apportionment to punish, for example, a clearly frivolous dispute, a non-cooperative Party, a particularly belligerent Party's representative and so on.

It is suggested that great care should be taken on punitively apportioning fees (and costs where permitted) between the Parties, particularly where reasons are being given. This can be relevant to fees where a head of claim asks for a decision of the Adjudicator awarding all of its fees to be paid by the other Party, and the adjudication procedures effectively require an Adjudicator to provide reasons for each decision on a head of claim. If there is evidence that anything other than what is allowed (e.g. the facts and law, or anything else expressly permitted) has been taken into account in apportionment, a decision might be appealed on the basis of excess of jurisdiction or partiality.

Putting adjudication on temporary hold

It is perfectly feasible that Parties may wish to either hold discussions at any stage of a contractual dispute procedure with a view to, say, achieving a settlement to a dispute outside that specific procedure (e.g. see the section on interim mediation by an Adjudicator in Chapter 14), or to have some time to take stock of where they are, say, in the light of differently presented information. A situation may therefore arise where both Parties agree during the Adjudicator selection and appointment process, that an adjudication should be temporarily halted. This is probably not a particularly important issue for a yet-to-be-appointed Adjudicator, or an appointed Adjudicator who has not received a referral document bundle signifying the commencement of the decision-making timetable. It is considered that the courts are unlikely to support a claim by one of the Parties later that a subsequent Adjudicator's decision should be rejected by reason of this joint agreement on the basis, say, that the Adjudicator selection process was disrupted and the appointment was thus not valid.

However, the situation is different if both Parties want an adjudication to be temporarily halted someway through the adjudication decision process itself. There is nothing expressly stated in the Construction Act or Scheme about this, nor is there usually anything within conditions of contract or contractual adjudication procedures. The most convenient solution would be for the situation to be anticipated pre-tender and incorporated within the contract or contractual adjudication procedures to permit a temporary hold to be placed on the adjudication process. This amendment would have to comply with the Act, or an option to use the Scheme could result.

The most practical procedure would probably be one that uses the existing adjudication procedures, particularly those relating to amending the adjudication timetable. Both Parties could agree with the Adjudicator to extend the adjudication timetable requirements in accordance with the procedures, noting that an Adjudicator may not be able to adjudicate on a dispute that would recommence at some indeterminate time in the future, due for instance, to other engagements. The solution appears to be for the Parties to estimate the time they need for their negotiations and to see if the Adjudicator will agree to an equivalent extension. If not, it would appear that the appointment of the Adjudicator would need to be revoked by the Parties and the adjudication process started again later, if possible with the same Adjudicator where that appears to be most beneficial to the Parties. If the Adjudicator agrees to the extension, the Parties can agree with the Adjudicator to put the decision-making process on hold and to recommence, if necessary, at the end of that holding period. As the Construction Act places no limit on the extensions of time which can be agreed by both Parties, and this will be the case in most contractual adjudications, there seems no reason why this process cannot be extended for as long as negotiations are taking place between the Parties.

Similarly, a Referring Party (it will be remembered that a Responding Party does not have that right under the Construction Act) can ask an Adjudicator for up to a 14-day extension, unilaterally or on behalf of both Parties. That time might well be adequate for many purposes, including settlement negotiations.

Investigations

An Adjudicator usually has total freedom as to the extent of the investigations that it can undertake. This can range from obtaining as much information as time permits to using only the written information provided by Parties. There is no right or wrong approach.

Information

Depending upon what, if anything, is stated in the adjudication procedures, an Adjudicator can 'request', 'require' or 'instruct on' issues such as the provision of further written information, copies of particular reports and documents and so on.

A time limit should be given for any receipt of information, for meetings with Party representatives, and for the compliance with the Adjudicator's requests and instructions whilst the adjudication is proceeding.

The Adjudicator can request that any legal submissions, information or other points arising or wanted to be drawn to the Adjudicator's attention are submitted only in writing, with a stated limited length, before, after or instead of a meeting, and that distribution is also, simultaneously, to the other Party.

Specialist assistance

Adjudication procedures usually permit an Adjudicator to obtain specialist legal and technical advice without the approval of the Parties, but with an obligation to advise the Parties that such instructions are being given. Besides being polite to the Parties who will ultimately have to pay for the advice, it is possible that the Parties might be able to suggest and agree with the Adjudicator the specialist(s) to be used. This would reduce the likelihood of a dispute by the Parties on the findings reached, or even enable them to agree there and then to settle the relevant specialist outstanding point(s) between them to save money on the use of specialists. Depending on how critical the specialist point is it might also, of course, lead to the Parties agreeing the dispute and stopping the adjudication process. Some adjudication procedures require that additional information such as the likely costs of specialist advice be provided. Whereas the Parties cannot use this information to prevent an Adjudicator proceeding to obtain the advice, it may be a catalyst for the Parties to agree the otherwise disputed particular specialist points, so negating the need for the specialist advice.

Another advantage to an Adjudicator of proceeding with the agreement of the Parties is that the Parties may agree to appoint specialists directly. This should avoid the necessity for the Adjudicator initially to pay the specialists' fees and expenses, or for the Adjudicator to be involved with any associated specialist reimbursement problems that might extend beyond the date of the Adjudicator's decision.

Specialist reports should be copied to the Parties, not least because they are paying for them, but also so that they are aware of information that an Adjudicator might be taking into account in reaching its decision. If time permits, the Parties should be given the opportunity to comment on the reports to the Adjudicator.

Site visits

Site visits are at the discretion of the Adjudicator, who should (assuming time is available to undertake the visit, which might be an overwhelming deciding factor itself) decide on the basis of whether the adjudication would be more economical, quicker and no less efficient than if no visit took place.

An Adjudicator might arrange a simple familiarisation visit to the site without any official Party representative(s) being present. On the other hand, an Adjudicator might consider it useful that one or both Parties should be given the opportunity to attend. The visit might be agreed to substitute for, or lead into, a discussion meeting with one or both of the Parties.

It will be noted that to have both Parties attend on site is unlikely to be a requirement of adjudication procedures, there being no significant difference between a site visit with one Party and holding meetings with individuals or single Parties, as long as information obtained that might be relied upon in a decision is communicated to an absent Party, to minimise the likelihood of an appeal on the basis of partiality or unfairness.

If one or both Parties for evidential clarification purposes specifically asks the Adjudicator to make a visit, it is suggested that if possible it is undertaken, unless the Adjudicator considers it clearly likely to be a waste of money and/or of limited time. Such a visit would, however, assist reassure the Party or Parties that the Adjudicator was attempting to properly understand the dispute, case or defence being presented, even if ultimately the Adjudicator decided to give little or no weight to what additional evidence, if any, was found. It has been suggested that where an Adjudicator declines a specifically requested site visit, this should be in writing giving reasons, to assist prevent giving cause for a later appeal. Generally, however, an Adjudicator can proceed as it thinks appropriate in the limits of the available timescale, and as long as it acts impartially, it does not have to give written reasons for everything it decides to do. It therefore seems unlikely that not giving written reasons for not having a site visit should create particular grounds for a later appeal, (whereas giving written reasons might do just that).

Meetings

As with site visits (and probably associated site meetings), an Adjudicator should consider seriously whether a decision can be arrived at more quickly, cheaply and just as effectively and impartially on a documents-only basis (bearing in mind that the documents provided are usually instructed to be a complete presentation of a Party's case), before deciding whether to have meetings with the Parties.

Adjudicator procedures will permit an Adjudicator to meet and question either or both of the Parties. This can be with Parties separately or together, but where separate, a formal record of relevant points that may be taken account of in the Adjudicator's decision must be given as soon as reasonably possible to the other Party. This is an obvious example of procedural fairness, as described in Chapter 11.

It is the Adjudicator's decision whether to have a meeting or not, where it should be held, who can attend and who the Adjudicator wishes to discuss what matters with. The date of any meeting should be one first and foremost to suit the Adjudicator and its timetable, not the convenience of a Party's representative. The meeting can be inquisitorial with the Adjudicator asking the questions. If there is lack of cooperation with the Adjudicator's requirements, the Adjudicator can terminate the meeting.

An Adjudicator is not usually obliged under adjudication procedures to agree to requests by a Party for a meeting, or for certain people to be present or to set the dates of meetings to suit the availability of Parties' representatives. It will be noted that the Scheme is more prescriptive than most adjudication procedures on this issue. For example, unless the Parties agree to the contrary, Paragraph 16 (1) of the Scheme permits any Party to be represented or assisted by advisors including legal advisors, but, unless the Adjudicator permits otherwise, Paragraph 16 (2) permits only one person to represent a Party when giving 'oral evidence or representations'. As with requested site meetings, where an Adjudicator declines a specifically requested meeting it has been suggested that this should be in writing giving reasons, to assist prevent giving cause for a later appeal. This suggestion is not necessarily agreed with, for the reasons given previously.

The submission of later additional information to the Adjudicator

Each Party should have a reasonable opportunity of putting its case and, if time permits, dealing with that of its opponent. However, a Party may supply substantial additional information to an Adjudicator later during the adjudication period. This might be information that is not new to the other Party, but will be new to the Adjudicator. ('Ambushing', where one Party supplies extensive documentation that is new to the other Party, is discussed separately earlier in this chapter.) Even if an Adjudicator is able to deal with it within the initial timetable before notifying its decision, there may not be time to receive or consider a response by the other Party.

This may not be a significant problem if an extension of time is agreed. Under the Construction Act and hence included in most standard adjudication procedures, a 14-day extension can be given by the Referring Party if the Adjudicator agrees. A Responding Party, for some unknown reason, does not normally in current standard procedures have such a unilateral option, (although, in fact, this could possibly be included as an enhancement to contractual adjudications which would still comply with the minimum requirements of the Construction Act). However, an Adjudicator may not have any available time immediately following the initial 28-day timetable period, due, perhaps, to prior commitments, and so may have to refuse to agree to the extension of time. It is possible that both Parties and Adjudicator might agree a longer extension if the Adjudicator only agrees to take the late information from the first Party into account if it results in the other Party having adequate time to review and respond to that information.

Should no extension of time be feasible because a Party or Adjudicator will not, or cannot, agree to one, the Adjudicator's options include either to not take the late information into account, or to accept it without giving one Party the chance to review it properly and respond. The latter carries a risk that it could give rise to an appeal against an adverse decision. The former is the most robust and impartial decision. The other alternative is to resign (*Balfour Beatty* v *Lambeth* [2002]).

If information is received later than requested (e.g. perhaps several days after an instructed deadline, but in practice in time to be taken into account) the Adjudicator must consider whether to accept or to refuse it in the light of the pertaining circumstances. For Scheme adjudications, Paragraph 17 must also be borne in mind. Whatever the selected course of action, the Adjudicator must remember that no matter how personally irritated it might be by the actions or inactions of one Party, it still has to act impartially (and possibly fairly in some adjudication procedures). The decision must be based as a minimum on the facts and relevant law (and any other factors specified in the adjudication procedures), otherwise there might be grounds for an appeal. The generally limited opportunities for financial penalties for dilatory or obstructive performance by a Party have been discussed elsewhere in this chapter with reference to non-cooperative Parties.

Information provided during an adjudication must normally be directly associated with the originally referred dispute. Occasionally it might be related to matters not initially referred to the Adjudicator. Even where permitted by an adjudication procedure, an Adjudicator is not normally obliged (i.e. it has the option) to include any matters a Party or the Parties only later agree they want included in the dispute. Once again, an Adjudicator might decide that it does not have the time, despite any agreed timetable extension, to deal with such matters or that they are more suitable to be addressed in a separate adjudication.

Cross-claims and counter-claims

A cross-claim (i.e. a set-off or an abatement) will normally be made by a Responding Party in response to a claim that money was owed, requiring that the cross-claim should be considered by an Adjudicator at the same time as making a decision on the referred dispute.

In litigation and arbitration, the various definitions and applications of cross-claims and counter-claims remains a hotly contested area and the same issues arise in adjudication.

The courts have made clear, however, (see "Payment provisions, withholding notices" later in this Chapter) that in adjudications that come within the Construction Act, for all categories of otherwise valid cross-claims or counter-claims, excepting possibly abatements, the correct notices must have been given before an Adjudicator can consider them. For projects outside the Construction Act, the correct notice provisions, if any, expressly stated in the contractual conditions of contract must be adhered to. These are possibly the overriding points to bear in mind in reading what follows, which is essentially:

- A description of what can constitute cross-claims and counter-claims.
- Various contractual implications that might effect whether cross-claims and counter-claims can be considered, even if accompanied by what appear to be valid notices.
- Other considerations for an Adjudicator eg available time.

A set-off is a cross-claim defence for money that can reduce or eliminate the original amount claimed by a Referring Party. For example, if there is defective work that others have been employed to rectify, the remedial works costs can be claimed as a set-off against the amount being claimed. If an agreed set-off is less than the original amount claimed, the original claimant need only pay the balance due. However, a Responding Party would not receive any money if the set-off was larger than the original amount claimed, as in that situation a separate claim must be made for the whole value claimed for the set-off. An Adjudicator and Party might also find it helpful to note that in litigation a successful set-off will result in the Party claiming the outstanding set-off debt being able to recover its costs from the original claiming Party, something not usually permitted in adjudication.

There are three most likely set-off circumstances arising in construction and certain rules can apply:

- 'Independent' set-offs, where the claim and set-off relate to different contracts. In this situation, the claim and set-off must both be liquidated (e.g. based on an invoice or certified) or readily ascertainable (e.g. by calculation from agreed rates).
- 'Transaction' set-offs, where both claim and set-off relate to the same contract. In this situation a set-off might relate to a debt or also to poor work or services, and so the set-off can be unliquidated (e.g. a reasonable estimate made in good faith of the effect of defective work or delay).
- 'Contractual' set-offs, where there is an express condition in the contract that, say, a certifier (such as a professional) must take into account any set-off that an Employer might be entitled to at the time of certification.

An abatement is not a set-off, but is another legally and separately identifiable type of cross-claim. Essentially, the Party against whom a claim is being made alleges that the claim is of lesser value because of, say, defects in the matter being provided, or because it is not fit for a stated purpose, or because it has not been completed. In other words, it is an amount of damages sustained as a consequence of a breach of contract. It will be recognised that certain types of set-off might be similar to an abatement. But if a cross-claim is stated and agreed to be an abatement, the specific difference is that whereas in some conditions of contract set-offs are expressly excluded from consideration, abatements usually are not. (However, notwithstanding any argument over whether a claimed abatement is an abatement or a set-off, an abatement might not be permitted anyway. For example, a certificate stated as being 'conclusive regarding the quality of the works' cannot have an abatement considered by an Adjudicator in a dispute on such a certificate. On the other hand, a certificate stated to be 'conclusive only as to cost' would not prevent such consideration.) An Adjudicator being given a cross-claim described as, and appearing to be, a valid abatement, cannot use a contract set-off exclusion clause as a reason for not dealing with it as it does not apply to abatements. Conversely, a Referring Party, finding that an Adjudicator has been presented with an abatement by a Responding Party, might attempt to argue that the abatement is, in fact, a set-off excluded by contract and hence from the remit of the Adjudicator! In this situation the Adjudicator must decide whether to proceed or not and on what basis. The Adjudicator can seek arguments from both sides, or seek external advice, perhaps in the latter two situations with an associated timetable extension.

A counter-claim will often be more clearly identifiable as a stand-alone issue capable of having its own separate adjudication. In litigation, there are practical advantages to this (depending on whether a Party is a claimant or defendant). For example, in the case of a set-off being claimed, in litigation only one Party will win its costs. With a counter-claim, because one or both Parties can be successful with their individual claims, in effect each Party might be awarded payment of its costs by the other (although in practice the courts might do a calculation that results in one Party only paying the balance of the two amounts). In adjudication, of course, apportionment of costs is not usually an issue for an Adjudicator. As for other cross-claims, the main practical problem for an Adjudicator (assuming both Parties agree to include the cross-claim even though not included in the initial referred dispute) will be having time available to deal with it.

The following is generally relevant:

- Cross-claims will be excluded if they do not comply with the requirements of the underlying conditions of contract, e.g. if there is an express Construction Act (or other) requirement that relevant notices and timing of notices must be given (see 'Payment provisions, withholding notices' in the next section of this chapter).
- The payment provisions of the Construction Act do not expressly limit any right to cross-claims, e.g. a main contract can still have set-off clauses, even those that allow set-off of amounts deemed due from other contracts, or which permit a Party to estimate the amounts to be withheld.
- Dealing with cross-claims against a disputed amount is not obligatory, but is permitted (if they are properly described in the dispute and have been the subject of proper Notices) by an Adjudicator if not expressly excluded by the pertaining adjudication procedures.
- Notwithstanding what is in an adjudication procedure, cross-claims against a disputed amount cannot be permitted by an Adjudicator if expressly excluded by the construction conditions of contract.
- If a cross-claim is apparently permitted in the construction contract, but relates to a debt on another contract, and where the primary adjudication dispute should be one 'arising out of a contract', it could be argued that the cross-claim should still be excluded.
- For non-compliance with contractual provisions to be a complete defence (i.e. a prerequisite before, say, money is paid or withheld), it has to be expressly stated in a contract to be so.
- Where there is no express statement, and thus there is possibly only a partial defence, this might mean repudiation of only a part of a claim to which any proof of loss applied.
- A separate adjudication on a dispute in which the cross-claim is the issue in dispute is perfectly acceptable, assuming it derives from within the contract between the Parties. (As noted, this would not apply where a cross-claim was a debt on another contract.)

- A cross-claim or counter-claim has to be presented within the timescale of the adjudication such that the Adjudicator feels it can deal with it (if it wants to) and in the time remaining available. If is not pleaded in (or as) the original dispute and even if both Parties are permitted to agree to widen the original dispute to include it, an Adjudicator may consider it needs the full 28 days to cover the initially claimed issue alone, bearing in mind that it will usually be possible to deal with a cross-claim, and certainly a counter-claim, as a separate adjudication.
- The grounds for a cross-claim or counter-claim should not be malicious or trivial, and this should bear on an Adjudicator's decision to deal with the issue or not.

Payment provisions, withholding notices

Non-adjudication Sections 110 and 111 of the Construction Act contain (and hence the underlying conditions of contract for projects falling under the Construction Act should also contain) certain notice requirements to be met before set-offs (but not necessarily abatements) can be applied. Before one Party can withhold payment that becomes due to another Party (or to deduct liquidated damages) it must provide notices containing specified information and further it must provide those notices by certain times. If either the notices do not contain the required information, or are not provided in time, due payment cannot be withheld. In addition, a withholding notice must be issued before the reference to adjudication (*VHE* v *RBSTB* [2000] and *Northern Developments* v *Nichol* [2000]). Notices must be in writing (*Strathmore* v *Greig* [2000]).

The courts (*Whiteways* v *Impresa Castelli* [2000]) stated that it was too late to raise an abatement in court for the first time as a defence to a money award in an Adjudicator's decision. They further stated that the Construction Act (in Section 111, which deals with the need to issue a notice of an intention to withhold payment from an amount claimed as due) makes no distinction between set-off and abatements. The courts then continued, stating that abatements do not need withholding notices. They do, however, need to be directly related to the disputed matters referred to the Adjudicator. For example, if a dispute specifically concerns matters referred to in a Valuation 10, it would be wrong for an Adjudicator 'to enquire into an alleged evaluation of Valuation 6 ... unless the notice of intention to withhold payment identified that as a matter of dispute'.

A further case supports this (*Miller* v *Nobles* [2001]). This was not an adjudication but an application for summary judgment for the payment of ten unpaid invoices, maintaining that the defendant was not entitled to withhold payment of the invoices because it had not issued a withholding notice. The defendant stated that it had made a previous overpayment to the claimant on an earlier valuation and that the sums now claimed were not sums due under the contract. The courts did not agree that account had to be taken of the previous overpayment (on the basis of *Whiteways* v *Impresa Castelli*, i.e. amounts could not be abated for matters outside what was described in that case as 'the four corners of the claim' unless those matters were mentioned in a withholding notice). The courts accepted that amounts could be reduced if their valuation was incorrect. Essentially they agreed that the amounts claimed could not be reduced by matters that should be, but were not, stated in withholding notices.

As explained in the previous section in this chapter, an abatement is an adjustment to a claimed amount to take into account that the amount includes for work that has not been completed or is sub-standard. Section 111 deals with withholding payment on amounts due, but as an abatement is an adjustment to an amount that is not totally due in the first place, it should not need a Section 111 Notice. Thus an Adjudicator, asked to determine an amount due (say, in an interim valuation), can legitimately assess what it considers to be the true value of the incomplete or sub-standard work without that notice.

A Scottish, case supports this (*S L Timber Systems* v *Carillion* [2001]) although this does not mean the English courts would necessarily also agree. An Adjudicator was asked to determine the sum due under a contract in an interim application but decided that the full amount applied for was due, with no deductions, because no Section 110 or 111 Notices had been issued. The courts were asked to consider the consequences of failing to serve Section 110 or a Section 111 withholding notices. The courts stated that there was no provision in the part of the Construction Act that deals with this issue that states that a failure to serve a Section 110 (2) Notice stopped a Party from claiming that a reduced amount was due. Further, Section 111 provides that a Party 'may not withhold payment after the final date for payment of a *sum due under the contract* unless he has given an effective notice of intention to withhold payment'. It was decided that the words 'sum due under the contract' was not the same as the sum claimed in the application and a Party was still entitled to ask the Adjudicator to value the works carried out under the application, i.e. to determine what sum was due under the contract. A dispute about whether the work in respect of which the application was made had been done, or had been properly measured or valued, could also be considered by the Adjudicator, even in the absence of a withholding notice. However, where the refusal to pay was on some separate ground, such as a right of retention in respect of a counter-claim, that would constitute an attempt to withhold a sum due under the contract and the Adjudicator could not consider it in the absence of a withholding notice. Essentially, the courts agreed that an Adjudicator could decide *an amount due* even if there are no valid withholding notices. However, any deduction from *an amount due* needs such a notice.

Incidentally, the courts also enforced the Adjudicator's award as they considered that the Adjudicator had acted within its jurisdiction when answering the question of whether one Party was entitled to full payment without deduction as a consequence of the lack of Section 110 and 111 Notices, even if it had answered that question wrongly.

Sometimes the words 'set-off' and 'abatement' are used interchangeably. Where 'set-off' clearly refers to an abatement, it can be treated as an abatement. This is not always understood by Adjudicators. In one case (*Woods Hardwick* v *Chiltern* [2000]), the courts considered that the Adjudicator had made a mistake in its decision by incorrectly deciding that Chiltern could not abate Woods Hardwick's fees, as no withholding notice had been made. Nevertheless, the courts supported this aspect of the decision, despite there being an error. (As already described, a similar error was made in the Scottish case of *S L Timber Systems* v *Carillion* and was also supported by the courts.)

Whether, other than for abatements, an Adjudicator is obliged to decide that anything claimed by, say, a Contractor, must be paid in full if, say, the Employer does not issue adequate withholding notices in time, is a matter of contract. For example, in some standard conditions of contract the amount due may be defined as the value of works currently executed (or similar wording), so that if a Contractor applies for an inflated amount in an interim valuation, an Adjudicator can still decide that payment of a lower value is more appropriate and apply an appropriate abatement. In other contracts (e.g. JCT with Contractor's Design, Clause 30.3.5) it has been argued (*C & B Scene* v *Isobars* [2001]) that without proper withholding notices, the amount due is whatever the Contractor is claiming, i.e. even abatements are not allowed. This has been supported by the courts (*VHE* v *RBSTB* [2000]). In the latter circumstances, an Adjudicator does not have the power to decide anything else, no matter how inflated the valuations might be.

Other points arising from cases dealing with payments and withholding notices are that without being mentioned in such a notice an Adjudicator cannot deduct liquidated and ascertained damages from the amount of an Adjudicator's decision if a claim has not been made at the time and in the manner required by the conditions of contract (*Nuttall* v *Sevenoaks* [2000]); sums can become due under the contract even if not certified (*Barr* v *Law Mining* [2001]); a withholding notice cannot be provided before the relevant application for payment to which the notice is intended to apply (*Strathmore* v *Greig* [2000]); an Adjudicator is not in 'excess of jurisdiction', merely in error, if it makes a deduction for a sum that should have been, but was not, the subject of a withholding notice (*KNS* v *Sindall* [2000]). (A problem that arises from the latter is where an Adjudicator does not realise that an abatement usually does not require a Section 111 Notice and as a result does not take abatement into account when determining an amount due. The courts would treat this merely as an error and support the erroneous decision. The implication is that, to be entirely safe, wherever possible include any claimed abatement in a valid withholding notice.)

Summary relating to abatements

The precise distinction between counter-claims, cross-claims and set-offs on the one hand and abatement on the other hand has always been a delicate question in both litigation and arbitration. It was, therefore, almost inevitable that it would become an issue in adjudication.

The approach taken by many practitioners has been that if the Notice of Adjudication asks an Adjudicator for a valuation of 'all sums due', then the Adjudicator may take into account and apply abatements whether or not they have been mentioned in a preceding Section 111 Notice.

Early court cases (which, it must be noted, only applied to projects falling under the Construction Act) did not appear to suggest that abatement was excluded. However, since then there appears to have been a hardening of judicial attitude and possibly a growing difference between Scotland and England.

In England, in a case decided on other issues (*Woods Hardwick* v *Chiltern* [2000]), the courts appeared to indicate in passing that there was no need to refer to abatement in a withholding notice before it could be taken into account by an Adjudicator in its decision. However, in a later case (*Whiteways* v *Impresa Castelli* [2000]) it was held that for the purposes of withholding notices there was no distinction between set-off and abatement and the latter had to be referred to in a valid Section 111 Notice before it could be taken into account. That view was confirmed by another decision (*Re A Company* [2001]) which makes it clear that it considered that it was the intention of Parliament that there shall be no entitlement to seek to rely upon any form of cross-claim or *abatement* without first specifying it in a Section 111 Notice.

In Scotland, however, a different approach is being adopted. In one case (*S L Timber Systems* v *Carillion* [2001]) the courts reviewed the various authorities but indicated that in Scotland it preferred the approach in *Woods Hardwick*. In other words, in Scotland (where, incidentally, it was also held that, unlike England, the entire claim must still be proved in the absence of a Section 111 Notice) an Adjudicator can still consider a dispute on *an amount due* which included whether claimed work had actually been done, or had been properly measured or valued. In other words, an Adjudicator could apply abatement in its decision, even where it has not been referred to in a Section 111 Notice.

There thus now appears to be a difference in practice regarding abatement between Scotland and England. It should also be noted that, at the time of writing, this subject continues to be the subject of court cases and the foregoing first instance cases may be overtaken by other decisions of equal weight and authority which may arrive at different conclusions. The issue may also, at some stage, be considered by a Court of Appeal.

The resultant position presents potential difficulties in practice for an Adjudicator. The foregoing and any subsequent decisions will no doubt be cited to the Adjudicator by both Parties and the Adjudicator in some situations may need specialist advice on the latest situation. However, what is interesting is that whether an Adjudicator includes or excludes abatement in its decision, and that is considered incorrect by a court, it may well be decided to be an error that falls within the Adjudicator's jurisdiction and the decision will be supported by the court.

To put matters in perspective, the majority of situations requiring a Section 111 (and Section 110) Notice will relate to interim applications and if something is not exactly right, it can usually be corrected in the next payment application. At worst, cross-claims can be made the subject of their own referrals.

Delays in copying submissions to the other Party

Under adjudication procedures a Referring Party must send a copy of the full statement of its case to the other Party at the same time as to the Adjudicator very shortly (e.g. within, say, 2 days) after the selection or appointment of the Adjudicator. Typically (e.g. Paragraphs 4.1 and 4.2 of the ICE Procedure), the date the Adjudicator actually receives the documents becomes the start date of the referral timetable (although this is not so clear in some adjudication procedures).

It is quite possible that one of the Parties, deliberately or otherwise, will delay sending copies of submissions to the Adjudicator to the other Party, thus reducing the time available for that other Party to respond. This applies particularly to the referral statement of case, as a delay in receiving the latter would have a direct impact on the time available for the Responding Party's principal response.

One situation that might arise would be if the Referring Party took longer than, say, a specified 2 days to send any, or a full set of, required documents to both Adjudicator and Responding Party, i.e. there is an equal delay to both. The Referring Party would thus be in breach of contract by being in breach of the adjudication requirements. If there was a short delay in excess of, say, the 2 days, equally applying to the Adjudicator and Responding Party, it is considered reasonable to merely delay the start (i.e. the date of the referral) to the date the documents were received, and not reduce the 28-day timetable. The Adjudicator would then proceed as normal. Nevertheless, as the delay will have been a breach of a specified contract procedural requirement (or in the case of the Scheme, a possible breach of a statutory requirement), the Responding Party might see this as grounds for an eventual appeal on the Adjudicator's decision. However, a robust Adjudicator could argue that the prime purpose of adjudication is to reach an impartial, rapid and inexpensive determination of a dispute (or whatever words might be used in the pertaining adjudication procedures) and to proceed as intended would best meet those objectives. Furthermore, the Adjudicator might argue that the Construction Act requires adjudication procedures to provide a timetable *with the object of* securing the appointment of an Adjudicator and referral within 7 days of the dispute. In this situation the procedures still do that, although they had not quite worked in practice, but there was no breach of the Act. This issue was discussed in the commentary following Section 108 (2)(b) in Chapter 2.

Another situation that might arise is one that resulted in a possible disadvantage being suffered by the Responding Party only. For example, if an Adjudicator received the referral documents promptly but, intentionally or otherwise, there was a delay in the Responding Party receiving them, with the Referring Party having no vested interest in extending the timetable. This could be (although an Adjudicator would not know it at that point) the first of a series of steady drip, a day here and a day there, 'spoiling' tactics, or unfortunate coincidences which might significantly impinge on the Responding Party's available time. In this situation there appears to be little an Adjudicator could do to reinstate the time 'lost' to the Responding Party, as any timetable extension needs the agreement of the Referring Party. This is unlikely to be provided if the latter is responsible for the delaying actions and they are deliberate. The generally limited possibilities of financial penalties that might be awarded against a deleterious Party have been discussed elsewhere in this chapter in the context of non-cooperative Parties.

Receipt by Adjudicator of without prejudice/confidential information

An Adjudicator accepting any information not provided to the other Party leaves itself open to claims of unfairness, partiality, not complying with natural justice and so on. An Adjudicator will therefore require all provided information (which to be safe must be assumed to be possibly relevant to making its decision, otherwise why provide it?) to be copied to the other Party.

However, a Party might send an Adjudicator so-called confidential information which the Party says is not to go to the other side, but which the Adjudicator only realises is the case after reading it. Although not an ideal solution (the obvious alternative being to resign), to be as impartial as possible it is suggested that the Adjudicator must write to both Parties saying what has happened, whether the information is considered relevant and, if so, that in view of the inability of the other Party to know the contents, the information will not be taken into account in the Adjudicator's decision. This is unlikely to satisfy the second Party, who will no doubt notify the Adjudicator that, if necessary, it intends to use the issue as a basis of an appeal because of the likelihood that such information did, in fact, influence the Adjudicator's decision.

Another situation that might arise is that a document might be sent to the Adjudicator headed for, example 'Without prejudice', dealing, say, with confidential discussions held between the Parties in an attempt to achieve a settlement. This document might be a single stand-alone submission; it may be deliberately or accidentally included within a referral or responding 'bundle'; it may or may not be copied to the other Party. It may not be headed as confidential at all, but might be confidential information included within a communication to the Adjudicator, only becoming apparent when the Adjudicator has read it or the situation is pointed out by the other Party. Once again, the Adjudicator must notify the Parties of the circumstances and ask if they both agree to waive confidentiality. If one Party does not do so, and the Adjudicator has read the contents, it will find itself in a similar situation to the previous one, i.e. notwithstanding that the Adjudicator might state that the information will not be taken into account in its decision, the second Party will probably notify the Adjudicator that it might use the issue as a basis of an appeal because of the likelihood that such information did, in fact, influence the Adjudicator's decision.

This is a difficult situation to deal with. The simplest solution may be for an Adjudicator to resign. However, if this is the automatic response expected in all similar situations, it might be expected that less scrupulous Parties, unsure of the strengths of their case, might deliberately create the circumstances for this to happen. Another alternative is for the Adjudicator to proceed regardless and hope that the decision will not be appealed. The Adjudicator might consider asking one of the Parties to seek an immediate court ruling on whether, under the particular pertaining circumstances, it is in order to proceed before actually doing so.

Receipt by Adjudicator of details of new pertinent case

An Adjudicator is obliged to take account of the law in reaching its decision. A situation might arise later in an ongoing adjudication where an Adjudicator receives information about a newly decided court case that might affect its decision. This might well be before the case is publicly reported and readily available to the other Parties, or one or both Parties might not be legally represented and expected to know the law, or one Party might bring it to the Adjudicator's attention in the belief that it supports that Party's case. What to do depends on the time available before a decision is required, but whatever else happens, if it is a case (and this applies equally to any other information) that an Adjudicator is likely to use in reaching its decision, the Parties must be told about it and, if possible in the remaining timescale for the decision, have the opportunity to respond to the Adjudicator.

The potential difficulty, as stated, is the available time, firstly, to communicate to the Parties, secondly, to receive their responses, if any, and thirdly, to read what has been supplied and take that into account where appropriate in the decision. With two weeks to go, say, before the decision, there is not usually likely to be a problem. However, if the new information only arrives with the Adjudicator a couple of days before the decision there clearly will be. There may be scope for an extension of time, but it is likely that a Party that perceives the latest information as likely to weaken its case will probably not be too helpful. If there is no extension of time, the answer, it is suggested, is that the Adjudicator must do its best in whatever time is available. It is then up to the Parties to decide if they want to appeal the decision (and risk incurring costs) on the basis, probably, of a supposed breach of natural justice. The courts might well take into account which Party would not agree a requested reasonable extension of time in deciding this.

Disagreements on meaning of Construction Act or adjudication clauses

In Chapter 10 on Jurisdiction, the scope of an Adjudicator's jurisdiction at the commencement of adjudication was considered under the headings of whether the adjudication provisions permitted, or both Parties agreed, that the Adjudicator had jurisdiction to decide certain issues, or whether there was no such statement or agreement and the Adjudicator decided for itself.

If a legal interpretation issue is in the initially referred dispute and the Parties agree to the Adjudicator's jurisdiction, an appointed Adjudicator must decide it, in the same way that it must decide any dispute once it accepts the appointment. However, there will be some situations where either the Parties do not agree that the matter is one for the Adjudicator to decide, or the Adjudicator declines to accept an appointment to decide the issue. In that case, the Parties must go elsewhere for a decision.

An example is a case (*Palmers* v *ABB Power* [1999]) where it was unclear whether scaffolding work on a power station was a construction operation under the Construction Act. In that case, the Parties went to the courts and obtained a quick decision. (As also stated in Chapter 10, it may be possible to obtain a quick jurisdiction decision from someone from a panel of Arbitrators set up by a professional body for just such a purpose.)

In another case, (*Homer Burgess* v *Chirex* [1999]), an Adjudicator made a unilateral decision that, in the circumstances of a dispute, pipework linking pieces of equipment was not 'plant' and hence the dispute was one concerning a construction contract and that it had jurisdiction. The courts disagreed. This illustrates the danger of an Adjudicator deciding on the interpretations of the Construction Act. The Parties and Adjudicator wasted considerable time and costs, which would have been prevented if the courts had been approached for clarification of the matter before the referral had taken place.

Another possibility is if an issue needing clarification arises which is not one initially referred to the Adjudicator, but becomes apparent later in the referral. If the Parties then agree during an adjudication that the Adjudicator has jurisdiction to decide an issue, it can (but is not obliged to) do so.

A different situation would be one where a Referring Party claims, after an Adjudication appointment has been accepted, that a contractual adjudication procedure paragraph does not comply with the Construction Act and wants the Adjudicator to use the provisions of the Scheme instead. This can be difficult for the Adjudicator, particularly where the procedures and/or the Adjudicator Agreement expressly state, as they sometimes do, that the Adjudicator is obliged to use the adjudication procedures stated in the contract, and the Adjudicator may have signed its Agreement confirming that it would do so. Further, the Parties, particularly the Referring Party, may also have signed the Adjudicator's Agreement by this time. Another potential problem is that the contractual Adjudicator appointment procedures may be different from those in the Scheme, thus giving rise to a possible objection by the Responding Party. The easiest solution might be for the Referring Party to issue a new Notice of Adjudication and start again. If the Referring Party did not want to do this, an alternative would be for the Adjudicator to note and record the arguments made by both sides and then continue on the basis of the initial adjudication procedures, leaving the Referring Party to take any further action it thought was appropriate. In practice, however, proceeding with an adjudication in these circumstances, perhaps now without the support of either Party for one reason or another, would probably lead to an Adjudicator resigning anyway.

The primary purpose of a construction Adjudicator's appointment will normally be to resolve a construction dispute, not interpret the requirements or associated meaning of the Construction Act, the Scheme or any other adjudication procedure. All standard adjudication procedures and statutory documents have been prepared with legal advice as to the validity of their provisions and any legal queries of this nature are primarily considered to be matters for the Parties to resolve. Unless an Adjudicator has agreed an initial appointment to deal with specific legal matters on the referred dispute, it is therefore suggested that such matters that later arise should be left alone for the Parties to deal with. If questions concerning legal interpretation arise, and they are not obviously frivolous, the suggested solution in most cases would be for the Adjudicator to suggest to the Parties that they temporarily halt the adjudication (as discussed earlier) and to seek a decision on the particular point(s) from the courts. An alternative would be for the Adjudicator to seek the specialist legal advice permitted by Construction Act-compliant adjudication procedures.

Another similar issue, on the face of it, is whether an Adjudicator can decide on the interpretation of conditions of contract clauses. In a case that did not impinge on the Adjudicator's jurisdiction (*Butler* v *Merewood Homes* [2000]), the courts decided that an Adjudicator could decide such matters in the same way as any other contractual dispute, and that it would support the decision, even if it was incorrect. Most disputes relating to the interpretation of the non-adjudication clauses of conditions of contract are unlikely to have Adjudicator jurisdiction implications and resulting decisions will similarly be supported by the courts.

Immediate responses to Parties

There may be occasions when attempts are made to place an Adjudicator under pressure to provide an immediate response to matters raised by Parties or their representatives. A typical situation is when an Adjudicator and the Parties meet face to face. Under most adjudication procedures this should only occur at a meeting (as discussed earlier in this chapter) arranged by the Adjudicator if it decides it wants one, to discuss matters the Adjudicator wants to discuss and with particular persons. In theory, it could be entirely inquisitorial with the only questions being asked by the Adjudicator.

A meeting, if required, will probably be some days (if not a couple of weeks to coincide with receiving a reply from the Responding Party) into the dispute. By that time the Adjudicator will have received and studied the Parties' submissions, refreshed its mind on the Construction Act and pertaining adjudication procedures, and there should be no lack of general preparedness. Even if difficult questions then arise, they should usually be avoidable by an Adjudicator stating that time is needed to study an issue before answering. Controversial matters can be requested to be put in writing and copied to the other Party for their response before replying. It should be remembered that a technical Adjudicator is not required to be an instant legal expert for instance, and vice versa, and that it would be unreasonable to expect an instant response in many circumstances.

Other than that, an Adjudicator will usually have time to give a question or statement a considered reply. This might take a few minutes to refer to a document such as the underlying construction contract, adjudication clauses, or reference notes. Occasionally some further study or advice may be necessary.

It is thus suggested that the solution includes having ready access to, and being up to date on, the contractual adjudication requirements and dispute as far as reasonably possible, requiring that, where necessary, questions should be in writing and copied to the other Party for their views, and then insisting on having the necessary time to study and prepare answers.

Umbrella Group Draft Guidance to Adjudicators using the Scheme

Unmanageable documentation

The issue

The Umbrella Group identify the issue in this area as being as follows:

> *Sometimes one Party submits unmanageable and disproportionate quantities of documentation to the Adjudicator, making it difficult for the other Party to respond and for the Adjudicator to reach his decision within the time limit.*
>
> *Some Adjudicators express concern that if they seek to limit the amount of paperwork that they receive or consider, they will be vulnerable to challenge either on the grounds of natural justice or because it would be a breach of Paragraph 17 of the Scheme (which requires the Adjudicator to consider any relevant information). They (wrongly) believe that they must take into account all information submitted to them in the course of the adjudication.*

The law

The Umbrella Group's view of the law is as follows:

> *As explained, natural justice requires (amongst other things) that the Adjudicator must give to each Party a reasonable opportunity to present its case and act fairly between the parties. However, 'fairness' must be set within the context of adjudication as a fast and interim procedure.*
>
> *Paragraph 17 of the Scheme requires the Adjudicator to consider 'any relevant information submitted to him'. Paragraph 13 gives the Adjudicator power to limit the length of written documents submitted to him.*

The guidance

The Umbrella Group suggest the following:

1. The responsibility for judging what is or is not 'relevant information' lies with the Adjudicator but in principle it is information that is evidential of the issues or events that a Party has to prove in order to further its case.
2. The duty in Paragraph 17 [of the Scheme] needs to be considered in the light of the powers given to the Adjudicator by Paragraph 13 [of the Scheme]. In particular, the Adjudicator may give directions as to the timetable and any deadlines or limits as to the length or quantity of written documents.
3. An Adjudicator could consider seeking (under Paragraph 19 (1) [of the Scheme]), consent to an extension of time in order to give the Adjudicator a proper opportunity to consider the papers and the other Party's response.

4. Remembering that it is in control of the adjudication procedure, an Adjudicator could at the outset consider limiting the amount of material to be submitted, taking into account the nature and value of the dispute. To prevent documents being 'drip-fed' to the Adjudicator, an Adjudicator should consider stipulating that all submissions and all documents must be received by a certain date.
5. An Adjudicator could consider requiring a Party to produce a concise statement of its case, cross-referenced to a bundle of back-up documentation, and a chronology (although an extension of time may also be needed).
6. An Adjudicator could consider whether the documentation properly relates to the dispute being adjudicated or to another issue (which may or may not be a dispute between the Parties).
7. An Adjudicator could consider using the power given to it under Paragraph 15 of the Scheme where a Party declines to comply with a request without showing sufficient cause.

Intimidatory tactics

The issue

The Umbrella Group identify the issue in this area as follows:

> *There is growing evidence that some Adjudicators are experiencing intimidatory or 'bullying' tactics from the Parties or their representatives during the course of the adjudication. For example, undue pressure may be put on the Adjudicator to adopt a course of action desired by one of the Parties. This may be by means of aggressive threats or other actions designed to reduce the control that the Adjudicator has over the process. It may involve:*
>
> - *making spurious challenges to the Adjudicator's jurisdiction*
> - *causing delay with the intention of obtaining an extension of time*
> - *deliberately confusing the Adjudicator through the use of technical or esoteric legal arguments*
> - *threatening to take no further part in the adjudication or to take legal action against the Adjudicator, or to report him to his professional institution.*

The law

The Umbrella Group identify the law in this area as follows:

> *Paragraph 13 of the Scheme lists some of the steps that an Adjudicator can take in order to determine the dispute, and Paragraph 14 requires the Parties to comply with any request or direction of the Adjudicator.*

The guidance

The Umbrella Group suggest the following:

1. Adjudicators should recognise such tactics early and counter them firmly but fairly. Adjudicators should remember that they are in control of the procedure.

2. Adjudicators, where necessary, could require that all arguments are put in writing with a copy sent to the other Party.
3. Adjudicators could consider refusing to discuss any matter over the telephone. If Adjudicators do discuss matters over the telephone they should keep a detailed record of any such telephone calls. It is suggested that great care be taken when communicating with one Party in the absence of the other Party. Where contact with one Party alone is unavoidable, it is suggested that the Adjudicator keep a detailed record of what is said and sends it to both Parties as soon as practicable. Where unsolicited telephone calls are received by the Adjudicator, it is suggested that the Adjudicator require that the information be conveyed in a form (for example, fax or e-mail) which can be sent to both the Adjudicator and the other Party.
4. An Adjudicator should respond politely but firmly if a Party is behaving in an intimidatory manner; it is also suggested that it may be appropriate, where arguments are being put forward aggressively or at undue length, to limit the length of the submissions that may be made out.
5. Adjudicators should always bear in mind that while issues raised by the Parties should be taken into account, the Parties or their representatives should not deflect the Adjudicator from the duties imposed upon the Adjudicator.
6. If one Party goes to court, for example to challenge jurisdiction, the Adjudicator should continue the adjudication unless both Parties agree otherwise or the court so directs.
7. The Adjudicator must remember that adjudication is a rapid procedure leading to a provisional decision and that, if need be, the complexities of the underlying dispute can be resolved in arbitration or litigation.
8. Adjudicators should not lose their tempers!

Chapter 13

The Decision

Introduction

At a certain point in an adjudication an Adjudicator will stop the information-gathering process in sufficient time (taking into account any other commitments) to consider, arrive at, produce and, if appropriate (e.g. subject to any requirement for payment of outstanding fees and expenses), issue its decision prior to the expiry of the pertaining time limit. That decision, provided as a formal written document, is the most important part of the adjudication.

An Adjudicator must use the facts, the applicable law and the relevant terms of the pertaining contract and adjudication procedures, to reach its decision. The decision should, in simple English, give the title of the contract; the names and contractual position of the Parties; the pertaining adjudication procedures and the underlying contract; the dispute to be decided with brief surrounding circumstances; how the Adjudicator appointment arose and pertinent details of what happened during the adjudication in terms of instructions given; resulting actions, documents and information and advice received; and finally, the decisions reached including fee (and, if relevant, costs) apportionment. Sometimes decisions and reasons on other issues such as interest will be provided, but usually only if requested. The pertaining contract clauses and adjudication procedures will usually need to be checked to see if such issues are mentioned, and if so, if they are expressly permitted or excluded. An example list of issues to be considered in a decision is provided in Appendix 5.

The courts have stated (*KNS* v *Sindall* [2000]) that an Adjudicator's role is to apply the terms of a contract in the same way as an architect, Engineer or surveyor, the difference being that the Adjudicator's decisions are more immediately enforceable. However, it is considered that in practice an Adjudicator will often arrive at a decision based on a 'balance of probabilities'. Adjudication procedures may sometimes also permit a decision additionally to be made on commercial or other grounds (but see Chapter 16 Insurance Implications). Either of the foregoing approaches might be at variance with, say, an ICE Engineer's decision, which will usually have been made in strict accordance with the requirements stated in conditions of contract.

Because of the limited time available to the Parties and the Adjudicator, it is suggested that an Adjudicator can accept any feasible factual allegation which has not been disputed. If they are disputed, the Adjudicator must weigh up the opposing factual evidence to reach its balance of probabilities decisions on those issues. It is considered that there is no obligation on a Referring Party absolutely to prove its case (the latter clearly not being required for decisions on commercial grounds, where permitted, anyway).

The Adjudicator must, however, reach a firm decision whereever possible, e.g. 'I consider that there is [no contractual entitlement/no additional money due to the Referring Party] under this subheading of the dispute'. The Adjudicator should not give a verdict simply as 'not proven' or 'not valid' (*Ballast* v *Burrell* [2001]). It will be appreciated that a decision, even if it is that no money is awarded or action is necessary, is still a positive decision.

It has been suggested that, in practice, an Adjudicator will often assess the merits of a dispute as it sees them from its own personal perspective as, say, a professional Consultant or site Contractor and then take more cognisance of the facts and applicable law that supports that view in order to reach its decision. Whilst realising that it is easier said than done, an Adjudicator must recognise this as a potential problem and try to ensure, as far as possible, that there is no related partiality in its thought processes that might affect its decision.

If a decision is appealed, the courts will be concerned that only relevant factors have been considered and that irrelevant factors have not in coming to a decision. The courts are not concerned with *how* factors are considered or the *weight* that has been attached to them by an Adjudicator. It will be noted, however, that it will be difficult for a Party to find grounds to appeal on such issues if no reasons are given.

This chapter deals with some typical decision-related issues. Express requirements are usually contained within adjudication procedures or relevant clauses of conditions of contract and, as stated earlier, those should be referred to for specific wordings.

Decision to be readily understandable

A primary concern of the Parties will be to receive a decision that is capable of being understood and complied with. Although there is generally no restriction upon the nature of the dispute that is referable to adjudication, in practice, most referrals and responses to them will be expressed ultimately in monetary terms. The decision itself will therefore normally be expressed the same way, involving whether or not there is considered to be an entitlement to money, and if so, the amount to be paid by one Party to the other. Expressing such decisions and the resulting actions required are thus usually straightforward.

There can be problems, however, even in monetary cases, particularly if the case being referred and the decision required of an Adjudicator are not adequately expressed. For example, in one case (*Cook* v *Shimizu* [2000]) matters relating to a Final Account were referred to adjudication. One Party apparently expected a decision expressed in monetary terms, whereas the other Party apparently expected the Adjudicator to decide a number of matters of principle. The actual decision issued by the Adjudicator recorded that 'all sums payable pursuant to the decision shall be paid by the Respondent to the Referring Party within seven days of the date hereof'. The Parties were unable to agree what that meant and resorted to litigation.

For non-monetary cases (e.g. a dispute on whether or not a wall has been properly constructed and, if so, what was needed to rectify it) a decision might require some very clear explanation of what, if anything, was considered by an Adjudicator to be wrong with the wall and the required feasible remedial measures.

Reasons

The pertaining adjudication procedures will usually state that an Adjudicator either may or must supply reasons for decisions reached. Sometimes the provision of reasons depends upon them being requested by one or both of the Parties.

Reasons should be factual as far as possible, not unnecessarily long, not argue the case of one of the Parties in any way that can be construed as biased, nor be emotive about the action or inactions of one of the Parties. This is notwithstanding any conflict (e.g. strong personality clash) that an Adjudicator might have had with either of the Parties.

Even when not a specific requirement, the provision of reasons is favoured by some Adjudicators, although sometimes they may be provided only as an attachment to the decision. They argue that reasons had to be prepared anyway in order to reach a decision and they might assist Parties understand and so accept the decision, so why not provide them, instead of just putting them away in a personal file in case they are needed later? Care must be taken. Firstly, to get them into a sufficiently polished state for detailed scrutiny by the Parties can take up a significant amount of the limited time available to produce the decision, and secondly, they can provide grounds for disputing the decision and result in a possible appeal.

Parties who want finality of decision would be against reasons. Otherwise, Parties might want them in order to identify reasons for appealing, rather than to expend additional costs later trying to obtain a more favourable decision through arbitration or litigation. Alternatively, reasons might be taken as an indication on how the dispute might be considered if it went to arbitration or litigation. They might thus influence whether to proceed with these or not. As a minimum, if the decision is to proceed, the reasons might provide assistance on how to improve the presentation of the case. Lawyers generally would like reasons in order to assist their Parties identify grounds for appealing, and insurers also want them, wherever possible, to assist distinguish insured from uninsured elements of the decision.

Sometimes adjudication procedures do not say when a request for reasons should be made (e.g. the Scheme. GC/Works is also confusing on this, as described in Chapter 5.) It should ideally be well before the decision is supplied, as the later the request the more likely that the decision could be delayed. A minimum of 14 days' notice would seem reasonable.

The provision of reasons will add to an Adjudicator's fees. Where the provision of reasons is optional, and only one Party requests them, then an Adjudicator would be justified in allocating its fees for this task to that Party.

Where no reasons are given for a decision and a set-off (or similar) is claimed, an Adjudicator should still make it clear if it has considered the set-off in its decision and state the decided amount due with and without set-off.

Otherwise, the Parties would not know if set-off still remained outstanding to be dealt with, perhaps, as a separate claim, later.

Allocation of costs and fees

The costs of an adjudication are the Parties' costs and the Adjudicator's fees and expenses, the latter possibly including the costs of obtaining any specialist assistance.

If an Adjudicator is considering awarding costs, possibly as a consequence of a request by one or both Parties, it should determine where its power to do so is coming from and if there are any limitations to the scope of that power.

The Construction Act and the original statutory Scheme adjudication provisions are silent on Party costs (although the proposed Scheme Amendments amend Paragraph 20 to expressly prevent an Adjudicator awarding such costs). Standard contractual adjudication procedures (e.g. ICE, CIC, JCT) usually stipulate that the Parties pay their own costs, but permit an Adjudicator to apportion its fees and expenses between the Parties. Others (e.g. GC/Works) permit an Adjudicator to award and apportion costs, fees and expenses. Ad hoc procedures might go either way, but some, particularly those between Contractors and sub-Contractors, may state that the Referring Party must pay the full costs, presumably as a deterrent to, say, sub-Contractors making any claims.

Regarding Adjudicator fees, it is common, although not necessarily correct, for Adjudicators to split them equally between the Parties, unless there is thought to be a particular reason not to (e.g. see the heading 'Frivolous disputes or a non-cooperative Party' in the previous chapter). It could be argued, quite reasonably, that fees should be apportioned in accordance with how much time was spent on each head of claim in a dispute and how the decision on each went. For example, if there is one item disputed and the decision is in complete agreement with what is claimed, then all fees should then be awarded against the Party refuting the claim. Whatever the answer, it is considered that apportionment is likely to be rough and ready.

There are several potential problems associated with an Adjudicator apportioning costs. For example, an Adjudicator should only apportion costs in ways, if any, expressly permitted by the adjudication procedures or the adjudication clauses of the conditions of contract, otherwise there is a risk of an appeal on the grounds of excess of jurisdiction. Similarly, an Adjudicator should not award costs against a Party just because it has found the Party's representative aggressive and awkward to deal with, otherwise there is a risk of an appeal on the grounds of partiality by a representative that might claim that although it robustly supported its client, it still did so within the boundaries of what should be allowed. (See also *Woods Hardwick* v *Chiltern* [2000].) Costs also cannot be awarded as damages in a situation where those damages are disputed (*Total M&E* v *ABB Building* [2002]).

Where costs *can* be awarded, they could form a major part of a claim and could overwhelm adjudication in terms of financial importance. This could affect the actions of the Parties and also impact on the Adjudicator, e.g. the Parties could be looking throughout an adjudication to criticise each other's conduct to assist influence the adjudication cost proportionment. This appears contrary to the spirit and intention of statutory adjudication, which should principally be aimed at solving a dispute cheaply, quickly and impartially.

In practice, complete cost details will probably not be provided to an Adjudicator in time for a proper assessment in the available timescale. Substantial costs could still be being incurred from after initial details had been submitted to the Adjudicator up to the time of receiving the decision, and beyond when studying the decision. In addition, from the time the costs are submitted, each Party might want to study and probably criticise each other's costs claims and submit those views to the Adjudicator, thus incurring more costs.

Another problem is that many Adjudicators will not be experts on the subject of costs, e.g. what is valid, whose costs can be claimed, from what moment of time and until when. Unless the Adjudicator can rely on the submissions and responses by the Parties, this may need to be the subject of outside specialist advice, which is likely to delay proceedings.

It thus appears often likely to be impractical for an Adjudicator to state in adequate detail the total costs allocation between the Parties within the Adjudicator's decision and within the timetable for that decision. It may be feasible for the Adjudicator to request costs up to a date well before the decision, apportion those costs, and then decide, say, that each Party should pay its own costs thereafter. Costs might well, however, be better decided as a later separate adjudication decision, or by arbitration, litigation or, say, separate expert determination. (It will be noted that a later contractual adjudication process might not be particularly straightforward if, for example, an ICE Engineer's-type decision process had to be gone through first before a defined dispute can be taken to have occurred.)

The courts have clarified some basic principles relating to the original Scheme, which, in some circumstances, might equally apply to other contractual adjudication provisions that are also silent on costs. For example, the County Court (*Cothliff* v *Allen Build* [1999]) decided that although not expressly permitted, the Construction Act and Scheme do not prevent an Adjudicator from awarding costs, where payment of costs was expressly requested by one of the Parties and arguments for and against had been presented to the Adjudicator. That part of an adjudication decision was then as binding as the rest of the decision. The County Court stated that if costs were not mentioned by the Parties, then an Adjudicator should not raise the issue itself (although if one Party raised it, the issue would come to the attention of the other Party anyway). A County Court judgment does not create precedent, but raising such decisions may be supported by a decision in a higher court. In fact, later, in another case (*Northern Developments* v *Nichol* [2000]), the courts decided that there was no implied term in the original Scheme to award costs, but that the Parties were free to agree amendments to the Scheme to permit this. Where both Parties ask for costs and do not suggest an Adjudicator has no power to award them, the courts will assume that the Adjudicator has been given jurisdiction by the Parties. The latter was confirmed by a case which occurred shortly afterwards (*Nolan Davis* v *Catton* [2000]), where it was agreed that if both Parties asked for costs, it could be assumed that the Adjudicator had been given jurisdiction to award them. It will be noted that the proposed Scheme Amendments, in Paragraph 20, prohibit an Adjudicator from considering Party costs.

Another case (*Bridgeway* v *Tolent* [2000]) dealt with contract provisions which made a Referring Party liable for the costs and expenses of both Parties and the Adjudicator. The courts decided that as the contract had been freely agreed to by both Parties, and as costs were not mentioned in the Construction Act, there were no public policy reasons why those provisions should not be supported.

In summary, unless expressly excluded, costs are either matters for the Parties (e.g. they each pay their own costs), they can contract otherwise, or if contractual adjudication provisions are silent on the subject they can both agree to let the Adjudicator decide.

Interest

If an Adjudicator is considering awarding interest, possibly as a consequence of a request by one or both Parties, as for costs and similar issues it should determine where its power to do so is coming from and if there are any limitations to the scope of that power.

Adjudication provisions may state that an Adjudicator can award interest (e.g. ICE, CIC). Some only permit this if included in the construction contract (e.g. JCT and possibly the Scheme. The latter is currently being interpreted both ways and needs clarification by the courts – see Chapter 7).

Notwithstanding whether it is permissible to award interest, in some circumstances it will be inapplicable anyway (e.g. a claim for money is lost) or an Adjudicator may decide not to award it (e.g. a claim for money may be won but the poor conduct of the winning Party during the adjudication might lead to an Adjudicator not including interest in its decision).

Where interest is awarded, a common current value is 8% simple on any amount awarded from when it should have been due up to the date of actual award. Should there be delays in payment thereafter (not a matter for an Adjudicator) the Parties will probably be guided by the Late Payment of Commercial Debts (Interest) Act 1998, which applies to debts owed by larger companies (more than 50 employees) to smaller companies (less than 50 employees). This Act entitles suppliers to charge interest at a statutory rate of 8% above base rate for the late payment of debts. Currently the base rate is around 6%. An Adjudicator, in practice, could ask for Parties' submissions on the clauses in the conditions of contract or adjudication procedures that might be applicable, their required interest, e.g. rate, simple or compound, when it should commence from, and even their calculations on the amount considered due. Submissions could be requested to include, where relevant, consideration of the statutory provisions of the Late Payment Act and associated Statutory Instruments.

Retention of decision until fees are paid

A refusal to pay, or at least a delay in paying, Adjudicator's fees is currently understood to be one of the biggest problems facing Adjudicators. Recovery can involve significant non-fee-earning management time, sometimes out of all proportion to the profit, if not the fee, that might be made by dealing with the adjudication itself. Adjudicators thus generally support the idea of withholding their decision until their fees have been paid. On the other hand, Parties or their representatives sometimes do not, perhaps wanting time to consider whether, in their opinion, an Adjudicator's hourly rate (in situations where that rate has not been agreed in advance) and the hours spent on reaching the decision, are reasonable. They are also uncomfortable in paying an Adjudicator in advance where there is a possibility of a complaint being raised about, say, an Adjudicator's jurisdiction or conduct of a reference, which might result in the Adjudicator forfeiting its entitlement to payment of its fees. However, there is nothing in the Construction Act regarding such a lien, Section 108 (2)(c) of the Act merely stating that the decision should be 'reached' within 28 days or any longer agreed period. The courts (e.g. *Bloor* v *Bowmer & Kirkland* [2000]) appear to support a more robust view, stating that a decision made within time but communicated out of time, was in fact out of time, unless there was an express or implied agreement by the Parties to an extension of time.

Some adjudication procedures require an Adjudicator to 'notify', or 'reach', or 'reach and notify' the decision within the timescale described in the Construction Act. Others clearly require the decision actually to be provided. Some (e.g. Paragraph 6.6 of the ICE Procedure) permit an Adjudicator (i.e. it is up to the Adjudicator to decide) to withhold the decision until all of its fees and expenses have been paid. Some (e.g. ICE Paragraphs 6.3/6.4) unilaterally provide the Adjudicator with an additional 7 days to 'reach', or 'reach and notify' a decision anyway. As has been discussed elsewhere in the text, no matter what is in the adjudication procedures or an Adjudicator's Agreement, there is no reason why an Adjudicator cannot propose that it retains its decision until its fees and expenses are paid. The Parties might both agree to this (although it is more likely that one Party might agree to pay the full amount due and then attempt to obtain the other Party's contribution). It could be argued that such provisions are consistent with the courts' support to Parties who are prepared to jointly accept a delayed decision (as for extensions of time in the previous paragraph). For the time being at least, it is suggested that the pertaining adjudication procedures and Adjudicator's Agreement should be referred to for guidance as to what might be acceptable in any particular circumstances.

Errors and their correction

The courts (e.g. *Bouygues* v *Dahl-Jensen* [1999]) will support a decision even if it contains errors in law, fact or calculation. That is not so critical when it relates to errors of reversible effect. For example, errors in decisions relating to payment and interim certificates can be corrected in later certificates so such errors are of a short-term nature anyway. Even errors in decisions relating to final payment and certificates can be resolved subsequently by arbitration, litigation or agreement between the Parties, as permitted by the Act. There is nothing in the Construction Act, the original statutory Scheme and most standard contractual adjudication procedures regarding the significant changing of a decision after it has been provided (although the currently proposed Scheme Amendments add a new Paragraph 22A to permit an Adjudicator to do so).

For more minor errors and slips, the courts (*Bloor* v *Bowmer & Kirkland* [2000]) have stated that for statutory adjudications 'In the absence of a specific agreement by the Parties to the contrary, there is to be implied into the agreement for adjudication the power of the Adjudicator to correct an error arising from an accidental error or omission or to clarify or remove any ambiguity in the decision which he has reached, provided this is done within a reasonable time and without prejudicing the other party', i.e. providing the contract does not state otherwise, an Adjudicator may correct a mathematical error or an ambiguity in reasonable time as long as it does not prejudice a Party. This finding was also confirmed in another case that was heard by the courts shortly afterwards (*Nuttall* v *Sevenoaks* [2000]).

Some adjudication procedures (e.g. Paragraph 6.9 of the ICE Procedure) do, in fact, expressly permit the correction of typographical mistakes and some errors and ambiguities in a decision, within a short period after reaching the decision. Occasionally, adjudication procedures can allocate even wider powers (e.g. the GC/Works/1 Procedure, by its association with the 1996 Arbitration Act, particularly Section 5.7 (see Chapter 5)).

Standard adjudication procedures generally do not, however, apparently permit an Adjudicator a change of mind, to insert omissions, rectify apparent Adjudicator carelessness, or correct any non-clerical mistakes. This could include, for example, not permitting an Adjudicator to change a decision if it found a previously supplied crucial document which the Adjudicator had misplaced, which would have changed the decision if taken into account earlier. As long as there is no 'bad faith', the Adjudicator would not be personally liable for such an initial omission, but it is suggested that a subsequent deliberate ignoring of a belatedly rediscovered crucial document and not drawing attention to it or to any consequences, might be construed that way.

An Adjudicator should clearly not expect to charge a fee for the time spent amending its errors. It might also expect to be under threat to reimburse fees and expenses already received if a decision would have been different if the error had not been made. For example, a now 'losing' and hence possibly aggrieved Party, particularly one who might have paid all (i.e. including the other Party's portion) of the Adjudicator's fees for the release of the decision, might feel that such fees should be refunded, particularly if it could prove that it had incurred costs as a consequence. In practice, an Adjudicator, aware of this possibility, might consider that, for example, it had done its best (albeit flawed) in the face of receiving many lever arch files of documents as part of a referral/response, that the decision probably was going to arbitration anyway, and perhaps the best practical course of action, from a personal point of view, was to remain quiet.

Peremptory compliance

A peremptory order is a 'do it or else' order, the 'or else' normally meaning the use of the courts for enforcement. In other words, an order for peremptory compliance enables the courts to instruct a Party to comply with the requirements of a decision, failure to do so giving rise to the consequences for breaching a court order, such as fines or imprisonment for contempt of court.

Although there is nothing in the Construction Act, adjudication procedures usually permit an Adjudicator (i.e. it is not obligatory) to order peremptory compliance with all or part of its decision (e.g. the Scheme, Paragraph 23 (1)). The reason for an order to be discretionary in adjudication procedures is that sometimes it may be decided that no money or action is due and thus there is nothing to enforce.

However, the courts (*Macob* v *Morrison* [1999]), in a case where an Adjudicator ordered one Party to comply peremptorily under paragraph 23 (1) of the Scheme with its decision that the Party should pay another Party some outstanding money, considered that:

- An adjudication peremptory decision was probably no more important than a non-peremptory adjudication decision.
- They might be reluctant to grant a mandatory injunction to enforce an adjudication decision involving *money*, as non-compliance carried the potential for contempt proceedings.
- A failure to pay money following an Adjudicator's decision was little different from not complying with a money judgment by the court and the sanctions should be no different, i.e. the usual procedure in money cases should be to issue court proceedings claiming the amount due, followed by an application for summary judgment.

There thus appears to be no value in an Adjudicator requiring peremptory compliance with a money decision.

Umbrella Group Draft Guidance to Adjudicators using the Scheme

Reasons in decision

The issue

The Umbrella Group identify the issue as follows:

> [Under Paragraph 22 of the current Scheme] *the Adjudicator is obliged to give reasons for his decision if requested by one of the Parties. The possibility of receiving a request at a late stage during the adjudication, or even after the decision has been given, has caused concern amongst Adjudicators.*

The law

Paragraph 22 of the current Scheme provides: 'If requested by one of the parties to the dispute, the Adjudicator shall provide reasons for his decision'. The Adjudicator also has a wide discretion, under paragraph 13(g) and (h) to 'give directions as to the timetable for the adjudication, any deadlines ... ' and 'issue other directions relating to the conduct of the adjudication'.

The guidance

The Umbrella Group suggest that for adjudications running under the Scheme, Adjudicators consider setting a date at the outset by which any request to give reasons must be made.

Current government intentions

Notwithstanding the possible changes to Paragraph 22 described in Chapter 8 dealing with the Draft Scheme Amendments, it is now understood that the government do not intend to change this area of the Scheme. This is perhaps because it is considered that the present Scheme provisions are adequate. It appears that the Umbrella Group draft guidance might remain unaffected in these circumstances.

The Parties' costs

The issue

The Umbrella Group identify the issue as follows:

> *There has been some uncertainty as to whether or not the Adjudicator has power to decide that one Party pay another Party's costs.*

The law

If agreed by Parliament, Paragraph 20 of the Scheme might be amended to make it clear that the Adjudicator does not have the power to decide that one Party pay another Party's costs. Paragraph 20 might provide:

> *The Adjudicator shall decide the matters in dispute. He may take into account any other matters which the Parties to the dispute agree ... but shall not take into account any matter relating to the legal or other costs of the Parties arising out of or in connection with the adjudication.*

Paragraph 25 of the Scheme, which states that the Adjudicator may determine who should pay his own fees and expenses, is currently thought likely to remain unchanged.

The guidance

The Umbrella Group suggest that [the changes in law should make it clear that] under the Scheme each Party will be responsible for their own costs, that the Adjudicator does not have the power to order one Party to pay another Party's costs and that it will not be possible for the Parties to give that power to the Adjudicator.

Current government intentions

Notwithstanding the possible changes to Paragraph 20 described in Chapter 8 dealing with the Draft Scheme Amendments, the government's intentions remain unclear in this area. This is perhaps because it is considered that case law now makes the position clearer. The foregoing Umbrella Group draft guidance is thus currently in abeyance.

Clerical mistakes and errors in the decision

The following guidance has been given as regards mistakes and errors.

The issue

The Umbrella Group identify the issue as:

> *Uncertainty regarding the extent to which an Adjudicator may correct accidental errors or omissions (errors) in his decision once it has been delivered to the Parties.*

The law

*Once the adjudicator has delivered its decision, its jurisdiction over the dispute is ended. However, where there is an error on the face of the decision, it seems that the Adjudicator retains a power to make corrections. As one judge (*Bloor Construction (UK) Ltd *v* Bowmer & Kirkland (London) Ltd *[2000] BLR 314, TCC) put it:*

> *'... in the absence of a specific agreement by the parties to the contrary, there is to be implied into the agreement for adjudication the power of the Adjudicator to correct an error arising from* ***an accidental error or omission or to clarify or remove any ambiguity in the decision*** *which he has reached, provided this is done within a reasonable time and without prejudicing the other party.'*

- *The power is contractual, so the parties are at liberty either to exclude the power or to limit it as they see fit (for example, in a contractual adjudication procedure). Alternatively, the Adjudicator or Adjudicator Nominating Body may set out the terms of the power in the appointment agreement.*
- *It is for the Adjudicator to decide whether there is an error or not.*
- *The following types of error are covered, according to* Bloor*;*
  accidental error;
  omission;
  clarification;
  removal of ambiguity.
  This generally reflects section 57 of the Arbitration Act and is wide ranging. It is clear, by analogy with arbitration, that an Adjudicator may correct his award to give true effect to his first thoughts and intentions, but may not change the substantive decision because he has second thoughts or intentions.
- *Correction of the error must take place within a reasonable time bearing in mind the speed of the adjudication procedure. In the case of* Bloor*, this was within two and a half hours of publication of the decision; however in another case, Mr Justice Dyson, as he then was, found it at least arguable that the Adjudicator had the right to correct a mistake after more than a week. (*Edmund Nuttall Ltd *v* Sevenoaks District Council *unreported 14th April 2000, TCC).*

The guidance

1. An Adjudicator must always check whether the Parties have made an agreement excluding or limiting its right to make corrections. If they have, the Adjudicator must comply strictly with such an agreement.
2. An Adjudicator may become aware that it has made an error, or one of the Parties may raise the matter. In either case, the Adjudicator must consider inviting Parties (or other Party) to make submissions to it.
3. When satisfied that an error has been made, the Adjudicator must correct it as soon as possible and the notify the Parties.
4. The Adjudicator should bear in mind how much time has elapsed since it delivered the decision, and any action that the Parties may have taken.
5. The Adjudicator has the right to decide that it has not made an error.
6. It may be appropriate (for example, in the case of decisions involving complex arithmetical calculations) to consider issuing a draft decision (in whole or part) a short time before delivering the final decision, inviting the Parties to identify accidental errors.
7. The Adjudicator's primary duty is to act fairly between the Parties.

Current government intentions

Notwithstanding the possible insertion of a new Paragraph 22A into the Scheme, as described in Chapter 8 dealing with the Draft Scheme Amendments, the government's intentions remain unclear in this area. The Umbrella Group draft guidance has been drafted to stand outside such amendments, if any, and to rely on case law. It should, therefore, be unaffected if amendments are not made.

Part III

Supplementary issues

Chapter 14

Miscellaneous issues

Introduction

This chapter considers some of the more general points arising out of adjudication, the detailed consideration of which does not fit readily into previous chapters. As always, reference should be made to relevant adjudication procedures and conditions of contract for actual procedures and wordings in each situation that arises.

What is a 'dispute'?

There has to be a 'dispute' otherwise there is nothing to refer for an Adjudicator's decision. But what exactly is a dispute? For the purposes of the following relevant points 'dispute' includes the phrase 'any difference' used in the Construction Act:

1. An Adjudicator has jurisdiction over disputes, not claims.
2. A dispute requires a claim and a rejection (*Monmouth CC* v *Costelloe & Kemple* (1965)), or at least a reasonable maximum time by which to respond, after which a claimant can then proceed with the dispute. There have been several other non-adjudication cases that state something similar.
3. The courts specifically considered what constituted a dispute in an adjudication case (*Fastrack* v *Morrison and Imregia* [2000]). They stated that:
   - A dispute arose once it 'had been brought to the attention of the opposing party and that party has had the opportunity of considering and admitting, modifying or rejecting' it. A rejection could be considered to have arisen where a Party refused to answer a claim. However, care is necessary. The courts decided in one Scheme case (*Griffin & Tomlinson* v *Midas* [2000]) that a Referring Party had failed to describe a dispute precisely enough. It consisted of several parts, and the courts stated that only one part had been sufficiently defined to identify it as a dispute falling within an Adjudicator's jurisdiction. The decision on the latter was enforced, but the Referring Party was held liable for the costs and expenses relating to the other less well-defined parts.
   - A dispute could also arise where there was 'a bare rejection of a claim to which there was no discernible answer in fact or in law.'

   From this it appears that a dispute occurs whenever a Party disagrees with any response to a claim by a second Party, or if there is no response within a reasonable time. How long a Party must wait if there is no response (or if a response is of the type 'We are still considering your claim submission' or similar) will presumably be a jurisdictional matter to be decided by the courts.

- A dispute could be in the form 'what sum is due?' without the need for any particularised or finalised amount being stated. (In this case, the Referring Party had partly particularised its claim, but widened it to include 'or any such other sums as the Adjudicator shall find payable').
- A 'single dispute' constitutes all issues at any one point in time which the claimant decides to refer to adjudication.

4. It follows that an admitted claim cannot be a dispute and hence cannot itself be subject to adjudication However, it is not unusual for a claim to be admitted but unpaid due to a disputed/unresolved set-off or counter-claim. There is nothing to prevent either Party referring a disputed set-off or counter-claim to adjudication, remembering that on an admitted claim subject to set-off or counter-claim, it is the latter which are disputed and thus should be referred to adjudication. (Cross-claims and abatements are dealt with in more detail in Chapter 12 Conduct of the Adjudication.)
5. The courts have also decided that even if a claim is indisputable, it can still form a dispute or difference (*Hayter* v *Nelson* [1990]), e.g. even if one person is indisputably right and one is indisputably wrong, there can still be a dispute between them.
6. A dispute can arise where a Party makes it clear that it definitely will not be paying a sum when it becomes due (*Discain* v *Opecprime* [2000]). Thus a dispute can occur even before the act of non-payment of an amount due.
7. A claim with no arguable defence can give rise to a dispute when the losing Party will not pay money that is due.
8. The Court of Appeal (*Halki Shipping Corporation* v *Sopex Oils* [1998]) has stated that 'dispute' had a wide meaning, including any claim notified to the other Party that the latter refused to admit or did not pay (whether or not there was a factual or legal answer to the claim).
9. The courts in another adjudication case (*Griffin & Tomlinson* v *Midas* [2000]) state that 'there must come a time when a dispute will arise, usually where a claim or assertion is rejected in clear language without the possibility of further discussion and such a rejection might conceivably be by way of an obvious and outright refusal to consider a particular claim at all'. They also state that there will only be a dispute where 'not only has there to be time to consider the claim or assertion', but also '... time to discuss and resolve it by agreement ...' and 'A dispute will not exist if the claiming party accepts or has no real answer to a justified criticism of the whole or part of a claim. Only when the stages of discussion or negotiation are at an end may there be a dispute which could be referred to Adjudication.'
10. What constitutes a single dispute has been considered by the courts (*McLean* v *Swansea* [2001]). It was decided in that case that various

issues such as valuations, provisional sums, extensions of time and loss and expense were all part of one dispute as they were all related to determining how much a contractor was entitled to.

11. The courts (Nuttall v Carter [2002]) undertook a detailed review of the relevant cases and other documents relating to what constitutes a "dispute". The judge said:

> *In my judgement, both the definitions in The Shorter Oxford Dictionary and the decisions to which I have been referred in which the question of what constitutes a 'dispute' has been considered have the common feature that for there to be a 'dispute' there must have been an opportunity for the protagonists each to consider the position adopted by the other and to formulate arguments of a reasoned kind. It may be that it can be said that there is a 'dispute' in a case in which a party has been afforded an opportunity to evaluate rationally the position of an opposite party* [but] *has either chosen not to avail himself of that opportunity or has refused to communicate the result of his evaluation.*
>
> *However, where a party has had an opportunity to consider the position of the opposite party and to formulate its arguments in relation to that position,* ***what constitutes a 'dispute' between the parties is not only a "claim" which has been rejected, if that is what the dispute is about, but the whole package of arguments advanced and facts relied upon by each side.*** *No doubt, for the purposes of a reference to adjudication under the 1996 Act or equivalent contractual provision, a party can refine its arguments and abandon points not thought to be meritorious without altering fundamentally the nature of the 'dispute' between them. However, what a party cannot do, in my judgement, is abandon wholesale facts previously relied upon or arguments previously advanced and contend that because the "claim" remains the same as that made previously, the 'dispute' is the same. The construction of the word 'dispute' for the purposes of the 1996 Act and equivalent contractual provisions, in my judgement, is not simply a matter of semantics, but a question of practical policy. It seems to me, that considerations of practical policy favour giving to the word 'dispute' the meaning which I have identified. The whole concept underlying adjudication is that the parties to an adjudication should first themselves have attempted to resolve their differences by open exchange of views and, if they are unable to, they should submit to an independent third party for decision the facts and arguments which they have previously rehearsed among themselves. If adjudication does not work in that way there is the risk of premature and unnecessary adjudications in cases in which, if only one party had had a proper opportunity to consider the arguments of the other, accommodation might have been possible. There is also the risk that a party to an adjudication might be ambushed by new arguments and assessments which have not featured in the 'dispute' up to that point but which might have persuaded the party facing them, if only he had had an opportunity to consider them. Although no doubt cheaper than litigation … adjudication is not necessarily cheap.*

In other words, and looking at what else was stated in this judgement, a dispute referred to an Adjudicator must only contain the information, evidence and arguments previously advanced between the Parties and supplied to each other up to the date of the notice of the intention to refer an issue to adjudication, ie the date of the Notice of Adjudication. Information that has not previously been made available by one Party to the other cannot be admitted or used by an Adjudicator to assist resolve the referred dispute. (Unless presumably, both Parties and the Adjudicator agree otherwise e.g. ICE Procedure Paragraph 5.2.)

Delay mechanisms in contractual adjudications

Adjudication procedures or conditions of contract may attempt to delay adjudication commencing until certain contractual procedures have been followed. The courts (*Carter* v *Nuttall* [2000]) decided that a bespoke pre-condition that mediation must take place prior to adjudication was not Construction Act-compliant, i.e. it prevented adjudication 'at any time'. Another case (*Mowlem* v *Hydra-Tight* [2000]) involved an NEC contract with a 'notice of dissatisfaction' clause similar to the ICE Conditions of Contract and a four-week negotiating period. Both the Parties and the courts considered this to be not Construction Act-compliant for the same 'at any time' reason.

It is suggested that the most important feature in both of these cases should have firstly been whether a dispute had occurred, as it is only if this is the case that the 'at any time' point needs to be considered. The ICE Conditions, Clause 66, give a definition, which again it is suggested, the Parties can, and do, contractually agree to. It is the validity of this ICE (and similar) clause that should be decided by the courts, but this has not yet occurred. If it should be accepted that by the time a matter of dissatisfaction has been referred to an Engineer, the legal definition of a dispute has already occurred, the delay awaiting the Engineer's decision is, without doubt non-compliant with the Construction Act's 'at any time' requirement. The option to use the Scheme therefore arrives before the matter of dissatisfaction is referred to the Engineer, and remains up to and beyond the time when the Engineer's decision is reached. This does not necessarily mean that the ICE Procedure is illegal. Both Parties (and if there are more than two, all Parties) are free to agree any contract and dispute resolution procedures, including any definition of a dispute, between them. (See also 'Order of Procedure for dispute resolution later in this chapter.)

The ICE Procedure becomes available to the Parties at the earliest when the Engineer's decision is reached. There are possible advantages to a Referring Party to want to wait to use the ICE Procedure. For example, the Referring Party might:

- Want a longer opportunity to prepare its case for either type of adjudication process so there is little to lose by awaiting an Engineer's decision

which might be in its favour anyway: this might apply even if the Referring Party's case is already complete.
- Consider the ICE Procedure to be potentially more advantageous to it than the Scheme.
- Not want to irritate the Client or Engineer on an ongoing project by precipitating adjudication before the Engineer had given its decision.

A further reason comes from Paragraph 5.3 of the ICE Procedure (see Chapter 3 dealing with the ICE Procedure), which permits a second Adjudicator to open up a previous Adjudicator's decision. This could be one resulting from an earlier Scheme adjudication on the same matter. A Referring Party can therefore have three bites of the cherry if the Engineer's decision is included, before needing to proceed to litigation or arbitration.

Perhaps what should be addressed is whether there is a need for an Engineer's decision anyway. Surely the days are long gone where the Parties believe that at one moment a professional can be an Employer's representative and by the switch of a title can become a completely impartial independent decision maker. It begs the question of what the value is of an Engineer who is most unlikely to meet any test for impartiality by which an Adjudicator would be judged (see Section 108 (2)(e) of the Construction Act), when there is an Adjudicator available who will act impartially?

Can an Adjudicator review all certificates and decisions?

The Construction Act does not confer specific decision or certificate open up, review and revise powers on an Adjudicator. Adjudication procedures will usually do so, but they may be limited in scope. For example, they may limit decisions to disputes on matters arising under the contract and not those that are outside it. Some decisions on matters arising *under* a contract will be stated as 'final and binding', 'final and conclusive' etc. and procedures may, in theory, permit an Adjudicator to review these. The Scheme, Paragraph 20 (a), for example, expressly excludes certificates, decisions etc. agreed in the underlying contract as final and binding. Others (e.g. ICE and CIC) are silent about exclusions, but at least one other (e.g. GC/Works, Commentary on Condition 60 dealing with the powers of an Arbitrator, such powers being also of an Adjudicator under Paragraph 6 of Condition 59 – see Chapter 5) apparently expressly includes even those that are final and binding. However, as a general principle, it is usually accepted by the courts that it is not within the public interest that a Party can reopen such issues.

Parties must also always be free *outside* a contract to negotiate a settlement at any time and to agree that the settlement will be final and binding. They can do this between themselves or use any third party they want, whether for an independent expert determination, mediation, conciliation, adjudication etc. as part of the Parties' attempts to settle outside a contractual framework. The latter external agreements (which are usually classed as contracts to pay money) are, it is suggested, outside any power of an Adjudicator 'to open up, review and revise matters arising out of or in connection with that contract.'

The courts (*Shepherd* v *Mecright* [2000]) have confirmed that an Adjudicator cannot deal with a full and final formally recorded and signed settlement between two Parties as:

(a) If there is a full and final settlement which is binding on the Parties, there is no dispute that can be referred to adjudication.
(b) The settlement was not a matter arising 'under' the contract, but arose 'in connection with' or 'out of' it. This will not always be a valid point, of course, because some adjudication procedures enable Adjudicators to deal with disputes wider than those arising 'under' a contract. Issues related to this are discussed in more detail under Section 108 (1) in Chapter 2.

The courts (*Finney* v *Vickers* [2001]) have, in fact, decided what appears to be a directly relevant issue. They considered that both Parties had reached a compromise settlement agreement on a dispute which incorporated an agreement not to pursue matters further by adjudication. The courts confirmed that the promise not to adjudicate was binding on the Parties.

As stated, some procedures (e.g. the Scheme) expressly exclude certificates, decisions etc. agreed in the underlying contract as final and binding. Also outside an Adjudicator's remit, except with the consent of the courts, are adjudications against companies in administration or administrative receivership.

The Construction Act permits adjudication decisions to be agreed as conclusive, under Section 108 (3). Any adjudication procedure that provides for this, or any default procedure that achieves the same effect (e.g. ICE Conditions of Contract Clause 66 (9)(b) or GC/Works Condition 60, Paragraph 1 where cross-reference is made to Condition 59 Adjudication) is thus in accord with the Act (even though they are contractual adjudication provisions) and it is considered most unlikely that the courts would permit such matters to be reopened by an Adjudicator.

Procedures might entirely exclude an Adjudicator from dealing with an issue already decided by itself or another Adjudicator within a contract (e.g. the Scheme, Paragraph 9 (2)). The courts (*Holt* v *Colt* [2001]) have considered a Scheme dispute where it was claimed that the dispute was the same as one previously decided. In that circumstance the Scheme precludes a further adjudication, i.e. an Adjudicator does not have jurisdiction to consider the same issue again. The courts decided that although the second dispute may have related to the same matter that arose in the first dispute, it did not relate to the first dispute. This is clearly a case-specific decision and other cases will need to be judged on their own facts. Some permit it (e.g. GC/Works, in which the Adjudicator is given the powers of the Arbitrator and Condition 60, dealing with arbitration, Paragraph 1, if certain notice requirements are met, lets an Arbitrator vary or overrule any previous decision of an Adjudicator). Others permit it only with the agreement of the Parties (e.g. ICE Procedure, Paragraph 5.3; CIC Procedure, Paragraph 26 in conjunction with Paragraph 30; JCT Procedure Paragraph 5.5 in conjunction with Paragraph 7.1). There are also some procedures that do not permit an Adjudicator to adjudicate on more than one dispute at the same time under the same or different conditions, without the consent of the Parties (eg see Scheme Paragraphs 8(1) and 8 (2)).

There are certain requirements that have to be met before a claim becomes a dispute and hence is able to be the subject of adjudication. This is discussed earlier in this chapter as well as in the detailed consideration of Section 108 (1) in Chapter 2.

Some conditions of contract require that certain procedures need to be followed before a dispute is deemed to have contractually arisen (e.g. ICE Conditions, Clause 66 (3)). Some conditions of contract may permit a previous contractual ADR procedure to be final and binding, or to become so by default if not referred to adjudication within a certain period (e.g. Conciliation under the ICE Conditions, Clause 66 (5)(b)).

The courts' position is clear. The House of Lords (*Beaufort* v *Ash* [1999]) has stated that the courts have the power to 'open up, review and revise' non-final and conclusive decisions or certificates, but not those that are expressly and clearly stated to be final and conclusive. The courts' jurisdiction was limited to deciding whether a conclusive certificate or decision was invalid because of bad faith or excess of power, otherwise they could merely enforce the agreement between the Parties. In addition, even though the courts cannot open a conclusive decision, if the Parties wanted an Arbitrator to do so, they must agree expressly that the Arbitrator and not the courts could do so. Otherwise, if the Parties agreed clearly in a contract that another person was to provide a binding interpretation of the Parties' contractual obligations, they would be bound by those obligations. The case concerned arbitration and JCT contracts which have no provision for making an architect's decision or certificate (other than the last certificate) conclusive, but commentators believe the House of Lords' decision applies more widely and would include adjudication decisions. The courts can thus review decisions that could be revised by subsequent arbitration or litigation or might involve an evolving situation (e.g. be a temporary decision on an extension of time or an amount payable on account), but otherwise they will support all final and conclusive decisions including those of an Adjudicator, except where derived in bad faith or excess of power.

It will be noted from the foregoing that, as for arbitration, adjudication procedures and conditions of contract can sometimes apparently permit an Adjudicator more powers than the courts. Whether the courts will always support this remains to be seen.

An Adjudicator should consider such issues when it studies the pertaining adjudication procedures, underlying conditions of contract and the history of any dispute resolution procedures that have led to the dispute then being referred to adjudication, before deciding how (or whether) to proceed further with an appointment. If an Adjudicator is unsure whether an issue is one that can be reopened, then it could ask for and assess the views of the Parties, review existing court decisions, obtain advice from others, including formal outside specialist advice, or suggest that the Parties obtain early clarification from the courts (*Palmers* v *ABB Power* [1999]). An Adjudicator must then decide for itself whether it is within its powers to review any particular issue and either withdraw or continue with the adjudication. It will be noted that it is the Referring Party's implied or express view that this is the correct position (bearing in mind that the Referring Party itself may have had legal advice on the strength of its position) so the Adjudicator might proceed letting both Parties deal with the issue in the courts after the decision had been made and if enforcement proceedings have commenced.

Interim Adjudicator decisions

The Construction Act, Paragraph 108 (3) states that all adjudication decisions are final unless changed by later agreement between the Parties, by arbitration or litigation (other than those, as has been suggested, that are agreed between the Parties to be final determinations). The Act does not preclude the later agreement between the Parties from being one arising as a consequence of further ADR process or adjudication. To that extent, initial 'final' decisions can therefore be considered to be interim decisions.

It will be noted that the Scottish version of the Scheme, Paragraph 21 (1) expressly permits interim decisions, e.g. an Adjudicator 'may make a decision on different aspects of the dispute at different times', whereas the nearest equivalent in the English version of the Scheme, Paragraph 20, does not.

In practice, a need for an interim decision may arise, for example, on a referral relating to an extension of time and/or additional costs, where those may need to be amended by later final determination, say, when a task is finally completed and more information is finally available.

If the initial adjudication decision had been agreed as conclusive (e.g. by agreement between the Parties or by following the Construction Act, Paragraph 108 (3)) or become conclusive by default (e.g. ICE Clause 66 (9) (b) or GC/Works Condition 60, Paragraph 1), it would probably not be possible for the Parties to revise it under the contract. Also, some adjudication procedures (e.g. Paragraph 9 (2) of the Scheme) can expressly preclude an Adjudicator acting on a dispute that had previously been decided in adjudication.

One way around such restrictions, might be for the Parties to agree to instruct an Adjudicator to act outside the contract, and the private agreement might be agreed to include using exactly the same adjudication procedures (and Adjudicator terms and conditions, if the same Adjudicator was involved in the review) as used for the initial adjudication proceedings. The Parties would then need to arrange to legally rescind the previous conclusive contractual agreement and substitute it with the later private one. This would not be a matter for the Adjudicator.

Another way to assist resolve potential difficult situations might be to use a specific descriptive form of words for each part of a continuing issue, so that each element is clearly separately identifiable. For example, on a series of extension of time disputes leading to a total extension of time, the reference could be 'to determine the extension of time, if any, from time "yyy" due to activity "zzz"', the criteria being stated differently in each case.

The most usual situation in building disputes is, however, where a contract has interim valuations followed by a final account application. Invariably, each interim valuation will be similar, if not identical in parts, to the preceding interim valuation. The last interim valuation will usually be even more similar to the final account valuation. In one Scheme case (*Sherwood & Casson* v *MacKenzie* [1999]), adjudication disputes were decided both on interim account valuations and a final account valuation. The decision on the final account dispute was appealed on the basis of it being essentially the same dispute previously decided. It will be noted that if this was the case, under Paragraph 9 (2) of the Scheme the Adjudicator was obliged to resign, i.e. it was a jurisdiction issue. The courts, in fact, decided that in this case, which involved a new loss and expense element plus additional documentary and revaluation support, there was a separate dispute (or rather, series of disputes) and hence they fell within jurisdiction.

Mixed Construction Act-compliant and non-compliant disputes

Situations sometimes arise where a single dispute requires decisions under several headings. Some of those headings might relate to operations that come under the Construction Act whilst others might not. How this is dealt with will differ between whether the Referring Party has chosen to exercise an option to use the statutory Scheme for deciding a dispute or whether it has decided to use the pertaining contractual adjudication procedure.

- Where the statutory Scheme is being used, only those heads of claim falling under the Construction Act can be decided.
- For contractual adjudication procedures whether or not they comply with the eight minimum requirements of the Construction Act, it is what is written within those procedures that dictates what heads of claim can be considered by an Adjudicator, not the Construction Act. In other words, notwithstanding whether there is a written construction contract, construction operations, work forming part of the land, a residential occupier situation etc., the contractual procedures apply to all heads of claim in the dispute. Hence, in a referred dispute where four heads of claim would fall within the scope of the Construction Act and two would not, all six will usually, depending on any express contractual procedure exclusions, fall within the jurisdiction of an Adjudicator appointed under those contractual procedures.

Following on from this, it is also suggested it could be very strongly argued that if the wording of the Scheme has been used to form contractual adjudication procedures, then that is what they are, and the latter situation should again apply.

Order of precedence for dispute resolution

As noted earlier in this text, the ICE Conditions of Contract (see Chapter 3) states a required order of precedence for dispute resolution so that, for example, a dispute as defined in those Conditions cannot be referred to adjudication until an Engineer's decision has been given (or a stated time with no decision has elapsed). As discussed earlier in the text, this pre-condition is not compliant with the Construction Act, where it is a requirement that a Party should be able to refer a dispute to adjudication complying with the Act *at any time*. This is reinforced by another case (*Carter* v *Nuttall* [2000]) where it was decided similarly that a mediation pre-condition was non-compliant. In a further case (*Herschel* v *Breen* [2000]), the courts decided that simultaneous adjudication and litigation was permitted, because the Act permits adjudication at any time. Thus, where a contract is a written construction contract for construction operations, as required by the Construction Act, a Referring Party has the option to use the Scheme (in whole or part) instead to resolve the dispute by adjudication at any time, i.e. out of sequence with any contractual sequence otherwise required for dispute resolution.

However, where the contract is not one to which the Construction Act applies, or the Referring Party does not choose to use its statutory rights under the Construction Act, the contractual obligations agreed between the Parties and relevant case law applicable to such situations will apply. For example, the courts will usually require Parties to follow any specific or implied order of dispute resolution procedures within conditions of contract. Also, the courts (*Enco* v *Zeus* [1991]) would not give summary judgment on the payment of a sum of certified money whilst an Engineer's decision remained outstanding which might have amended the amount of money in dispute. In another case, the courts (*Cott* v *Barber* [1997]) stated that they would stop a litigation action where there was arbitration or an ADR provision in the contract. In other words, it would appear that if there is a specific dispute resolution process in a construction contract (which might be in several stages such as conciliation, adjudication, arbitration), a Party can request the courts to stop another Party side-stepping these by going straight to litigation.

An interesting point to consider occurs with regard to precedence and to Paragraphs 1.1 and 2.1 of the ICE Procedure. Paragraph 1.1 states that the ICE Procedure prevails over the ICE Conditions of Contract in the event of there being differences between them. Paragraph 2.1 confirms that a dispute can be referred at any time. It could, therefore, be argued that if the intention of the drafters was to have the ICE procedure comply with the minimum requirements of the Construction Act, the dispute being referred to the ICE procedure is one defined by legal precedent as a dispute, not by the ICE Conditions of Contract. Hence under Paragraph 1.1 this is what must be considered by an Adjudicator. There is therefore no conflict between the ICE Procedure as written (which supercedes the ICE Conditions on this point) and the minimum requirements of the Act.

A Referring Party can therefore wait until an Engineer's decision has been made, but that is up to the Party – it does not have to. Using the 'legal precedent' definition of a dispute, such a dispute can be referred at any time. It could thus be argued that as the ICE Procedure is Act-compliant, there is therefore no option for a Referring Party to use the statutory Scheme, just an option to bypass the normal contractual order of precedence given in the ICE Conditions of Contract for resolving a contractual issue arising between the Parties!

Interim mediation by an Adjudicator

A situation arose where an appointed Adjudicator was asked by the Parties to act as a Mediator in private discussions between those Parties, the mediation failed and the Adjudicator recommenced the adjudication and made a decision. The courts decided (*Glencot* v *Barrett* (2001)) that despite both Parties agreeing to the Adjudicator's involvement in the mediation process, there was the possibility that matters that arose or impressions gained at private (or 'caucus') mediation meetings could give rise to 'unconscious or insidious bias' and have influenced the Adjudicator's later adjudication decision. There was no criticism of the Adjudicator, who had asked for the views of both Parties as well as obtaining legal advice before continuing with the adjudication.

Problems arose, of course, firstly, because the mediation was unsuccessful and secondly, because the same person acted as Adjudicator and Mediator. There are, however, various ways that both mediation and adjudication might be undertaken by the same person but with a reduced risk of appeal against an adjudication decision. They include making sure that both Parties are aware of the potential problems and having them agree in writing that, despite the risk of unconscious or insidious bias, the adjudication process can recommence in the event of mediation breakdown.

It might also be possible (although this is usually a key element of the process) to agree not to have private caucus meetings in the mediation. If the latter are to remain, ground rules need to be established for what mediation information can be taken into account by the Adjudicator in subsequent recommended adjudication, up to permitting the Adjudicator to use for its adjudication decision any relevant confidential information or impressions it obtained from the mediation caucuses. As the latter should normally remain confidential to each Party, there would need to be a clear written waiver of the need for the Adjudicator to be unconsciously unbiased, to be impartial etc. (see Chapter 11). It will be noted that the latter would be a waiver or change in a statutory right. To have any chance of being supported by the courts it would need to be agreed for the specific dispute, not a general contract agreement between the Parties to operate in this manner should such a dispute arise during a contract.

Decisions of irreversible effect

Other than where the Parties have specifically agreed to accept a decision as a final determination under Section 108 (3) of the Construction Act, under that Act an Adjudicator's decision should be reversible by litigation, arbitration or agreement between the Parties at a later stage. It is therefore possible that some non-money disputes could be subject to successful appeal, because of the unalterable effect of enforcing a decision.

Examples where the effect of Adjudicator's decisions can be permanent even if not expressly agreed to be so, include:

- A Party obliged to act as a result of an Adjudicator's decision may find in later arbitration or legal proceedings that the decision is overturned. The Party incurs irrecoverable loss and/or delay in putting things back to how they should have been. It might also incur irrecoverable costs if it had accelerated the Works following an adjudication decision that was later reversed.
- Irreversible actions taken on site (e.g. concrete incorporated into foundations or structures demolished and built over), with associated costs and time.

- Where a Contractor exercises a suspension right due to an Employer's non-payment of money following an Adjudicator's decision, both Parties may incur irrecoverable costs and/or delay, including those relating to restarting, if the decision is not supported by subsequent arbitration.
- One Party gives another Party money as the result of an adjudication decision, the dispute is referred to arbitration with the opposite decision being reached, but (a) meanwhile the receiving Party has gone into receivership (see later in this chapter) or is insolvent and so cannot repay money initially awarded or (b) there is no mechanism for interest to cover the temporary loss of use of the money.
- A dispute concerns whether an Employer can determine a Contractor's contract (e.g. for poor performance), the Employer obtains a successful decision and determines the contract, but an Arbitrator reverses it later.

From the foregoing it is apparent that adjudication decisions can have a significant effect on Parties and/or the Works.

It is suggested generally that where an Adjudicator is asked to become involved in disputes on technical specification, design or other issues that are likely to give rise to irreversible decisions, the possible implications should be pointed out to the Parties. If the Referring Party wishes to continue then so be it. It is, after all, a matter between the Parties, not the Adjudicator, and such issues are 'disputes' which come under the auspices of the adjudication process.

An Adjudicator should also avoid being involved where any claim is against a company in administration or administrative receivership, without first ensuring that proceeding with the adjudication has the consent of the courts. This is because Section 11 (3) of the Insolvency Act 1986 prevents legal proceedings from being pursued against a company in administration or administrative receivership without the consent of the courts. The courts (*Straume* v *Bradlor* [1999]) decided that adjudication also comes under Section 11 (3).

Potential insolvency of a Party, trustee stakeholder accounts

A typical Adjudicator decision is one where an Adjudicator awards one Party money to be paid by a second Party, as soon as possible. The losing Party may know that the dispute will go to arbitration and that the decision might be reversed, but on occasions might be concerned that it will be difficult to have the money returned, not least because of possible insolvency of the first Party.

In a case under a CIC Procedure (*Bouygues* v *Dahl-Jensen* [1999]), the Referring Party, Dahl-Jensen, went into voluntary liquidation before an Adjudicator's decision was reached in its favour. The courts supported Dahl-Jensen's request for summary enforcement even though the decision might be reversed in later litigation or arbitration in favour of the other Party and, because of insolvency of Dahl-Jensen, it would not be able to recoup its losses. This decision was appealed. The Court of Appeal stated that, although normally a counter-claim was not a defence to the enforcement of a decision, where there was insolvency involved, the Insolvency Act 1986 provided for a mutual set-off, i.e. the original claim and any cross-claims or other latent claims merged into one claim. In such circumstances it would be inappropriate to support summary enforcement of the original claim alone. In this case, no notifications of cross-claims were made at the original court appeal and so the Court of Appeal supported the original court decision to enforce the Adjudicator's award.

Another case (*Herschel* v *Breen* [2000]) confirmed that any request that an Adjudicator's decision should be stayed because of doubts that the initial winning Party would be able to repay if the decision was later overturned must be supported by strong evidence (e.g. demonstrating financial difficulties) for the courts to consider it. In a further case (*Rainford* v *Cadogan* [2001]) the courts stated that they would not normally enforce summarily an Adjudicator's decision where there was serious doubt about an ability to repay money should subsequent litigation or arbitration reverse the earlier decision. In this case, as the Party that might have had to pay back money was in administrative receivership, the courts decided that there was a reasonable chance of a financial recovery, and supported the decision.

On a similar point in yet another case (*Parke* v *Fenton Gretton* [2000]), a statutory demand was made to enforce payment by a Party of an amount decided by an Adjudicator. It was noted that there was a valid cross-claim in existence that was the subject of other ongoing proceedings. The courts decided that it was not right to enforce the Adjudicator's decision if that might make the Party concerned bankrupt before that Party had the opportunity to benefit from a favourable result from the ongoing cross-claim proceedings. The statutory demand was therefore set aside.

A concerned Party might request an Adjudicator in its decision to order the placing of any award money into a separate 'ring-fenced' account, pending a later award being given which might overturn the initial adjudication decision. Such an order by an Adjudicator has already been supported by the courts (*Drake & Skull* v *McLaughlin & Harvey* [1992]). This was a pre-Construction Act case where the courts permitted an Adjudicator to order the placing of disputed set-off money into a trustee stakeholder account.

In anticipation of such a request being made (which would delay payment on an Adjudicator's decision), stakeholder accounts can be expressly excluded under a contract. This option may no longer be viable anyway. In a post-Construction Act case (*Allied London* v *Riverbrae* [1999]), one Party was unsure that the second, Referring, Party was financially sound and asked for the Adjudicator to order disputed money to be placed in a deposit account in the joint names of the Parties. The Adjudicator decided the Construction Act did not give him the power to instruct this. The courts decided that, although the Adjudicator might in fact have had a duty to consider the suggested joint deposit account proposal, delaying payment by any means was contrary to the purpose of the Construction Act and an Adjudicator in such circumstances could not logically or lawfully make any order that had that effect.

This is supported by Section 111 (4) of the Construction Act. This requires that, notwithstanding whether an 'effective' notice to withhold set-off has been given, if an Adjudicator makes a monetary decision that includes any of the original amount initially referred to adjudication (which might exclude the claimed set-off in part or totally), the awarded amount must be paid at the latest of (a) 7 days from the decision or (b) the final date of payment, as if the 'effective' notice had not been given. In other words, where an Adjudicator's decision includes that the whole or part of an initially disputed amount must be paid, that decision must be complied with not later than the latest of (a) 7 days from the date of that decision or (b) the final date of payment if there had been no set-off. This allows no scope for the delay that would otherwise result if the original amount, being disputed by the Party claiming outstanding set-off, had to be put into an account pending the result of a later decision in arbitration or litigation.

It is relevant to note that the courts (*Allied London* v *Riverbrae*) also decided that, even if the Adjudicator had considered the stakeholder account issue further, there had not been any significant supporting evidence provided to support the claim of uncertainty concerning financial standing. If there had been, and actual or potential insolvency became an issue, it is suggested that in the context of current court decisions the result may have been different. As discussed earlier, the courts will not automatically enforce payment on an Adjudicator's decision where there is a strong argument that a Party may not be able to recover money initially paid, if subsequent litigation or arbitration produced an opposite decision to that provided originally by the Adjudicator. (However, it will also be noted that if the late receipt of money could be shown to be seriously detrimental to the Party initially awarded it, the situation might be different again.) One way to provide protection in such a situation would be to pay the money relating to the adjudication into the courts (or even a trustee stakeholder account!) This would surely have had the support of the courts in these particular circumstances. However, it will be noted that such a scenario does not warrant having a mandatory trustee stakeholder account clause in adjudication procedures, as this would be inconsistent with the intention of the Construction Act and the need for courts to robustly support valid Adjudicator decisions and the transfer of money due as soon as possible.

Legal 'disclosure' of adjudication documents

In arbitration and litigation, certain relevant documents held by each Party must be notified and can be disclosed to the other Party.

Normally precluded from disclosure (i.e. 'privileged'), unless a Party wants to disclose them, are documents between the Parties and their legal advisors and documentation, statements and information being undertaken primarily in anticipation of litigation or arbitration. Also excluded would be documentation and other records relating to without prejudice meetings and discussions that were attempts to reach a pre-arbitration or pre-litigation settlement. (It will be noted that merely entitling documents as 'privileged' or 'without prejudice' does not make them so, and sometimes the advice of the courts is sought to resolve disputes on this issue.)

It is assumed that if a decided dispute is referred later to arbitration or litigation in accordance with Section 108 (3) of the Construction Act, the foregoing rules will also generally apply. For example, for statutory and contractual adjudication all documents and evidence produced by a Party to an Adjudicator which should have been copied to the other Party anyway to support its case or defence, and everything produced by an Adjudicator, including specialist outside advice and its decision, would be disclosable, but not the precluded documents listed in the preceding paragraph.

Where an adjudication had taken place outside the contract with the agreement of the Parties in an attempt to reach a private settlement, it is suggested that documents produced entirely as a result of that adjudication would be precluded, unless both Parties agree otherwise.

Such matters should not be for the Adjudicator.

Health and safety

It is possible that an adjudication decision may unintentionally involve instructions to do something that is, in practice, unsafe, or leave in place something unsafe. However, for non-monetary disputes, an Adjudicator should normally be deciding, for example, whether work complies with the contract and not what should be done as a consequence. An Adjudicator should have indemnities in place, so will have limited exposure (see the commentary on Section 108 (4) in Chapter 2 on an Adjudicator's liabilities and indemnity requirements), although other professionals engaged by the Parties might have liabilities.

For example, where an Employer's monitoring professional believes something arising from an adjudication decision to be unsafe, that professional should instruct the situation to be remedied and a Contractor should expect to be reimbursed, if appropriate, in accordance with the relevant clauses in the conditions of contract.

If a professional engineer working for a Contractor's believes work is unsafe, but an Employer's monitoring professional will not intervene with an instruction, the Contractor's engineer must still act the same way as when finding any other unsafe practice on site. In other words, the Engineer should comply with its professional obligations and statutory health and safety requirements (e.g. give instructions personally to make the situation safe, notify upper management to give those instructions if necessary, failing that inform the relevant safety officer, and then the local Health and Safety Executive representative). If no action is even then taken (which is hopefully most unlikely), professional ethics may require the Contractor's engineer to resign, but it is suggested that this must be a personal decision based on the facts of the individual situation.

The Construction Act gives a Contractor a right to suspend performance in the case of non-payment, and this could include non-compliance with an adjudication money decision. This could, in theory, lead to leaving the site whilst it was in an unsafe condition. This should not happen, as a Contractor has health and safety obligations, whether the site is being abandoned as a result of the Construction Act or any other reason, e.g. it must be left safe for everyone, including third parties.

Limitation periods

Although, under the Construction Act, a dispute can be referred to adjudication for resolution at any time, an Adjudicator must apply the relevant law in reaching a decision. Legal limitation periods in contract (and adjudication will usually be a contractual procedure) are normally six years from the end of a contract, except where the contract is signed as a deed in which case 12 years applies. There would thus appear to be a sound case for an Adjudicator to decline to deal with any referrals arising after those dates. In practice, disputes are mostly going to occur during a contract, its maintenance period or, at the latest, whilst attempting to agree a final account, so very late referrals are most unlikely to arise.

Joining of disputes on related issues

The use of the same Adjudicator on related disputes, although perhaps using different contract forms (e.g. a main contract and a sub-contract, or a main contract and a Professional Services Agreement, should prevent obtaining different decisions on what is an underlying common dispute. Adjudication procedures (e.g. the Scheme Paragraph 8 (2)) and the ICE Procedure Paragraph 5.7 will usually permit this, but only if the Parties and Adjudicator agree.

There are, however, various factors the Adjudicator must bear in mind and draw to the Parties' attentions to attempt to resolve, it being assumed that the latter are in at least initial basic agreement with the idea of joining the disputes. For example, an Employer as a Referring Party might not agree to extend the timescale to assist, say, a Referring or Responding sub-Contractor it is not in contract with. Also, separate contracts may have different adjudication procedures that may not be easy to reconcile (e.g. the need for reasons to be given, a restriction on revising final certificates, awarding costs and so on, might be different between the procedures). There may be new potential conflicts of interest to be investigated and Adjudicator terms and conditions to be reconciled.

Because of this, the joining of related disputes, particularly ones where the contracts concerned are not from the same 'family' of contracts/sub-contracts, which is also likely to be the case where, for instance, a Professional Services Agreement is involved, may be less likely to be agreed than perhaps is envisaged.

Contract (Rights of Third Parties) Act 1999

Essentially a person who is not a Party to a contract may, in his or her own right, enforce a term of a contract if (a) a contract expressly permits it and (b) the term purports to confer a benefit on the third party.

Section 8 of the Construction Act enables third parties to take advantage of contractual arbitration clauses. Possibly based on this, where there are express adjudication provisions in the contract, it has been argued that third parties also have adjudication rights. This will no doubt be tested in the courts when such a situation arises.

Third party rights can be excluded by expressly stating that the Parties do not intend a term or terms to be enforceable by a third party, e.g.:

> *A person who is not a Party to this contract shall have no right under the Contracts (Rights of Third Parties) Act 1999 to enforce any of its terms, except and to the extent that this contract provides for the Act to apply to any of these terms.*

It will be noted that similar wording such as this might be applicable for use by an Adjudicator in its Agreement with the Parties.

Pre-action protocol

In October 2000, a pre-action protocol was introduced in England, applicable to almost all construction disputes. The protocol describes the steps that must be taken by both Parties before proceedings can be issued in court. It does not, however, apply to proceedings for the enforcement of an Adjudicator's decision.

Chapter 15

Appeals and enforcement

Introduction

As has been described in the preceding text, there have been, and continue to be, a significant number of referrals of adjudication decisions to the courts. The referrals are primarily firstly, by a 'winning' Party seeking the courts' support in making a 'losing' Party comply with an adjudication decision and secondly, by the 'losing' Party seeking a court decision that an adjudication decision is invalid for some reason and hence it does not need to be complied with. Frequently, those opposing positions are made by the Parties at the same court hearing.

Scope for a Party to avoid adjudication and/or appeal against enforcement

There is significant scope in theory for a Party to attempt to avoid adjudication or to appeal to the courts against a decision on the basis of a claimed invalidity of an Adjudicator's decision.

In practice, although a significant number of appeals against adjudication decisions have been decided by the courts, it is apparent that the scope for a successful appeal is limited. Generally, as discussed earlier in this text, the courts will support robustly any adjudication decision, except where the Adjudicator can be shown to lack, or to have exceeded, its jurisdiction or, less commonly, to have acted partially or in bad faith. There is also a possibility to appeal on the basis of not observing natural justice, although where the line is drawn on this is not always totally clear. Lack of jurisdiction may be due to there being no 'dispute' at the time of the referral; for statutory adjudication a 'construction contract' coming into force before the Construction Act came into operation or the dispute not being related to a 'construction operation'; an Adjudicator not being properly selected; an Adjudicator deciding on an issue not referred to it or one it had no power to deal with and so on. In addition, although an Adjudicator can make errors of fact or law, (e.g. *Bouygues* v *Dahl-Jensen* [1999] and *C&B Scene Design* v *Isobars* (Court of Appeal [2001])) errors of law that go to its jurisdiction can lead to a successful appeal. Bad faith and similar issues have been discussed earlier in Chapter 11 on Procedural Fairness and under Section 108 (4) of Chapter 2 dealing with the Construction Act.

It is considered that the following issues might be usefully addressed, often because of actual court decisions, as being possible grounds for appeal in different circumstances:

- A 'dispute', whether as defined in a contract or by case law, had not, in fact, arisen.
- There was no 'construction contract' as defined by the Construction Act and therefore the option for a Referring Party to use the Scheme did not exist.
- The dispute was not one 'arising under' the contract, (or whatever was stated in the pertaining adjudication procedures).
- Applied adjudication procedures did not comply with the Construction Act and the Adjudicator, with no authority to do so, had insisted, although the Referring Party had not opted to use them, that the Scheme provisions should have been applied instead.
- An Adjudicator's appointment did not comply with the adjudication procedures in some way. Express wording in the adjudication procedures or underlying contract may, however, have overcome this (e.g. Paragraph 10 of the Scheme which states: '*Where any party ... objects to the appointment ... that objection shall not invalidate the ... appointment or any decision he may reach*').
- There is a lack of impartiality, e.g. a possibility of bias due to an Adjudicator having a financial involvement with one of the Parties or behaving unacceptably to one of them (as discussed in Chapter 11).
- An Adjudicator had acted in 'bad faith' (as discussed earlier in Paragraph 108 (4) of Chapter 2).
- An Adjudicator had acted with 'excess of power' (e.g. deciding on an issue not referred to it).
- An Adjudicator was guilty of 'misconduct', say, by not enabling each Party to have a reasonable opportunity to put its case and deal with that of its opponent. Misconduct would not necessarily include a failure by an Adjudicator to address all the arguments presented (*Miller* v *National Rivers Authority* [1997]). Such an apparent or actual failure might reasonably be due to a too late submission of information, issues considered to be of no weight or being vexatious, malicious or trivial, or ignored because of not being copied to the other Party. The courts (*Macob* v *Morrison* [1999]) have also shown that they would not support invalidity arguments based on the short timetable provisions of the Construction Act.
- An Adjudicator had considered matters not forming part of the referred dispute.
- An Adjudicator had taken into account in its decision matters not brought to the attention of one or both of the Parties.
- There was a 'manifest error' on the face of the decision. The courts (*Dixons* v *Murray-Oboynski* [1997]) have stated that a manifest error in a decision is an error that is evident on its face, some examples of which are included elsewhere on this list e..g., it could be an act or omission of an Adjudicator, dealing with issues not requested to be addressed, considering evidence not made available to one or both Parties and so on.

Essentially, a manifest error has to be seen, e.g. that the task given to an Adjudicator had not been performed. This would not be easy (even impossible) to prove if little information (e.g. no reasons) were given. However, it should be noted that where an error is related to an error of law, fact or calculation (perhaps including one that, if corrected, would alter a decision), the courts (*Bouygues* v *Dahl-Jensen* [1999 and 2000]) have indicated that they would support such an Adjudicator's decision. They will not do so, however, where the error of law means that the decision became one an Adjudicator otherwise had no jurisdiction to make (e.g. *S L Timber Systems* v *Carillion* [2001] and *C & B Scene Concept Design* v *Isobars* [2001]). The courts will also not support a procedural error resulting in an Adjudicator making a decision outside its jurisdiction (*Barr* v *Law Mining* [2001]).

- An Adjudicator decided on a matter previously decided by the same or another Adjudicator, or one that is 'final and binding', where such matters were excluded from consideration by an Adjudicator, perhaps because of the pertaining adjudication procedures.
- An adjudication decision was supplied after the time limit specified, (but it will be noted that some adjudication procedures permit this).
- An Adjudicator was wrong not to deal with a cross-claim in its decision. (It will be noted that in an appeal the court would only be considering the effect of not dealing with the cross-claim on the validity or otherwise of the decision. The courts would not decide on an amount to be used to offset the amount decided by the Adjudicator in its decision.)
- Summary enforcement of a peremptory decision (or one where the adjudication procedures specify implementation 'without delay', such as in the ICE Procedure) should not have been ordered whilst a relevant 'triable' issue within the decision (such as a counter-claim or set-off), which there is a bona fide intention to have determined by way of litigation or arbitration, remained outstanding (*Enco* v *Zeus* [1991]).
- Adjudication not complying with the rules of natural justice as far as it was reasonably possible under any constraints imposed by the Construction Act and pertaining adjudication procedures (see Chapter 11).
- Other than where the Parties have specifically agreed to accept a decision as a final determination (e.g. under Section 108 (3) of the Construction Act), an Adjudicator's decision should be reversible by litigation, arbitration or agreement between the Parties at a later stage. It might be possible to appeal against some non-money dispute decisions because of the unalterable effect of enforcing the decision. However, the courts may not be too sympathetic. For example, the courts (*Bouygues* v *Dahl-Jensen* [1999]) supported a decision in favour of paying money to a sub-Contractor in voluntary liquidation, even though the decision itself could have been later reversed and there might be no chance of recovering any money paid over. (Also of relevance to the latter is that Section 11 (3) of the Insolvency Act 1986 prevents legal proceedings from being pursued against a company in administration or administrative receivership without the consent of the courts, so such a decision might not ever be reviewed anyway.)

In practice, it is likely that one or more of the foregoing (or other) objections will be raised with an Adjudicator during (and usually at an early stage of) an adjudication process, probably to be used by a Party (most likely the Responding Party) only if the Adjudicator's decision is an adverse one for that Party. An Adjudicator should initially obtain the view of the other Party, and check if there is anything in the pertaining adjudication procedures or underlying conditions of contract which might clarify the issue raised. If not, an Adjudicator can review existing court decisions; obtain advice from others, including obtaining formal outside specialist advice; suggest that the Parties obtain early clarification from the courts (*Palmers* v *ABB Power* [1999]); decide for itself the correct decision; or concur with the Referring Party's view (bearing in mind that they may themselves have had legal advice on the strength of their position) and let both Parties deal with the issue at enforcement proceedings in the courts after the adjudication decision had been made.

It will be realised that matters raised by the Parties in court after an adjudication decision has been reached will be outside the control of an Adjudicator.

Enforcement

There have been a number of referrals to the courts. Generally, but not always, they have supported the Adjudicator concerned and decided that its decision is valid. The following briefly describe a number of situations that arose and the courts' decisions.

Firstly, the courts (e.g. *Macob* v *Morrison* [1999]) decided that, notwithstanding that a contract contains arbitration clauses, an Adjudicator's decision is to be considered binding (until finally determined by legal proceedings, arbitration or agreement between the Parties in accordance with the Construction Act) and can be enforced by summary judgment through the courts.

Further, the courts decided that a decision remains a decision under the Construction Act which can be enforced, even though procedures might not have been 'fair', a notice of dispute had been given after the contract had been terminated, or a decision was wrong on a point of law, fact or calculation. (See *Macob* v *Morrison* [1999], *A & D Maintenance* v *Pagehurst* [1999] and *Bouygues* v *Dahl-Jensen* [1999], respectively.) In practice, it appears that this robust (and rapid) approach is resulting in most adjudication enforcement proceedings currently taking place through the courts.

Where a dispute is not complex, there are two main ways the courts can assist a Party obtain a quick settlement of issues awarded in an Adjudicator's decision if a 'losing' Party does not comply as quickly as will usually be implied or specified in the adjudication procedure and/or the decision itself.

- The courts (*Macob* v *Morrison* [1999]) considered that the usual procedure following an adjudication decision, *particularly in money cases,* would be for a Party to ignore any peremptory order procedure even if the option exists, and just issue court proceedings claiming the amount due, followed by an application to the courts for summary judgment, under Order 14 of the Rules of the Supreme Court, all in accordance with normal debt recovery procedures. This procedure could also be enhanced by a Party requesting a court to reduce the timetable for normal debt recovery (*Outwing* v *Randell* [1999]).

- For non-monetary cases, it has been suggested that the correct course of action is to apply for a mandatory injunction.

The courts (*Herschel* v *Breen* [2000]) have decided that where proceedings on the same issue are being pursued in parallel, say in court or arbitration, and a decision is imminent, they might not enforce an Adjudicator's decision.

The courts also might not support a decision where there is a real possibility that in the event of the Adjudicator's decision being overturned by later litigation and arbitration, it may not be possible to recover any money paid under that earlier decision due to, say, the financial status or insolvency of the initial 'winning' Party. The courts in an early adjudication decision (*Drake & Scull Engineering* v *McLaughlin & Harvey* [1992]) used Order 14 proceedings to force a main contractor to comply with a pre-Construction Act Adjudicator's decision to place disputed set-off money into a trustee stakeholder account. (See also 'Decisions of irreversible effect' and 'Potential insolvency of a Party, trustee stakeholder accounts' in Chapter 14.)

If a 'winning' Party can satisfy the courts that at a later trial on the same issue it would be awarded a substantial sum, but not necessarily the amount actually being claimed, in a situation where the courts might not otherwise support a decision, they may instead order an interim payment under Order 29.

In some cases (e.g. *Crown House* v *AMEC* [1989]), the courts decided that the matters were too complex (e.g. there were important legal 'triable' or cross-claim issues remaining to be resolved) to award even an interim sum and the Parties needed to sort out those matters more fully in court over a longer period (*Lathom* v *Cross* [1999] and *Atlas Ceiling* v *Crowngate* [2000]).

The Construction Act and the Scheme do not prevent a Party from making cross-claims including set-offs and hence a Party can attempt to resist an enforcement of the Adjudicator's decision by claiming set-off of amounts due to it. (See also 'Payment provisions, withholding notices' and 'Cross-claims and counter-claims' in Chapter 12.) Some commentators have suggested that it might have been preferable, to meet the intention of adjudication to provide a rapid decision, for the Construction Act to have expressly stated that 'insofar as money decisions are concerned – the Adjudicator's decision shall be enforceable as a debt free of set-off or counter-claims'. Unfortunately it does not.

Where no reasons are given within an adjudication decision and a cross-claim has been made, an Adjudicator should still make it clear whether it has considered that cross-claim in its decision. If this has been done, the courts can use Order 14 summary judgment procedures to enforce the decision, as there is no unresolved relevant 'triable' issue. The courts (*Tubeworkers* v *Tilbury* (1985)) will not, in practice, allow a cross-claim to be raised to defeat an Order 14 application where Parties have expressly agreed to exclude a right of set-off. The courts (*Gilbert Ash* v *Modern Engineering* [1974]) require clear words to exclude that right.

Chapter 16

Insurance implications

Introduction

All Parties engaged on construction operations have insurance cover for their activities and liabilities. Some insurances are obligatory by statute (for example, Employers Liability 'EL' liability cover to protect against their liabilities to their employees) or by virtue of the requirements of the conditions of contract (for example, Contract Works and Third Party 'PL' liability cover). Other insurances (for example, Contractor's Plant and Equipment) may be taken out voluntarily. It is usually a requirement for professionals, and also Contractors with staff performing similar duties, to have Professional Indemnity 'PI' cover for liabilities due to any negligent act, error or omission arising from a breach of professional duty. There are also other possible insurances, including cover for delays in completion/loss of revenue, shipping risks, transport, car fleet, damage to employer's existing property and occasionally latent defects.

Insurances may be on an individual Party annual renewal basis, but larger and/or complex projects commonly have a project insurance package covering all or most of the Parties for a comprehensive range of risks. Most of these insurances will be subject to an upper limit and a lower excess, i.e. the Insured is responsible for amounts outside the insured limits.

The major potential impact of adjudication on Insurers is that should an Adjudicator with a short, say, 28 day timetable a money decision against one of their Insureds, a decision that had to be complied with very quickly, the Insured would seek a rapid contribution from their Insurers.

Briefly, Insurers are naturally concerned that they should have some involvement in disputes which result in them having to pay out money, that the dispute should be properly researched over an appropriate period of time, and that their legal liability for and the amount of relevant payment should be clearly defined in the decision, with the possibility of appealing the decision before paying out any money. These issues will be discussed in more detail shortly.

Before continuing, it is useful to try to establish some perspective. For example:

- Research covering around 600 disputes occurring in the first two years of operation of the Construction Act indicated that under 5% of those disputes involved an Employer and a professional. Although there was a wide range, the average value of disputes involving money was less than £120,000. Actual amounts awarded, if subject to insurance reimbursement, would often have been subject to an excess which would have reduced that amount.

- For ICE Employer/Contractor type money disputes, there is little effective difference in consequence between an adjudication and an Engineer's decision, except that an Adjudicator may be perceived as being more impartial and, importantly, adjudication decisions are rapidly enforceable. Engineer's decisions do not seem to have given Insurers significant problems in the past.
- Some adjudication procedures, such as GC/Works in the current text, permit commercial decisions, and Insurers who only cover legal liability risks became concerned that they might inadvertently reimburse Insureds for things they were not insured for (see later). However, such adjudication procedures are not the most used of those procedures available and so their impact is currently small.

Problem policies

The main policies that might create problems for Insurers are said to be:

- Professional indemnity 'PI' cover for professionals covering their liability to their client and to third parties, and also those undertaking similar tasks in Contractors' offices.
- Public Liability 'PL' cover to third party property and personal injury.

Insurers' concerns

Insurers' concerns have been stated to include:

- An adjudication decision which may result in a monetary insurance claim is made very quickly on what may be a complex issue that would previously have taken much longer to sort out, with no opportunity for Insurers to take a significant degree of control.
- There is a risk that an Adjudicator, although meant to act impartially, does in fact have some connection with the Party claiming from the Insured. This may particularly be so where that Party is permitted to choose an Adjudicator who is pre-named in the contract as from a list included in the contract.
- The Adjudicator might be inappropriate for the dispute concerned, e.g. an Engineer Adjudicator on a cost dispute or a Quantity Surveyor Adjudicator on a technical dispute.
- There is no comeback on the Adjudicator if its decision is wrong/overturned by arbitration or the courts, and Insurers had meanwhile paid out on a claim, which might have resulted in unrecoverable money (e.g. a new structure had been provided or the insured client had gone into liquidation), or at least lost interest through temporary lack of use of the money.
- Whether generally an adjudication decision gives rise to a legal liability. It has been argued that it does, as (at least some) adjudication has a statutory

basis, and most adjudications are now enforceable by arbitration or the courts. In addition, the courts (*Straume* v *Bradlor* [1999]) have also confirmed that adjudication was a quasi-legal process falling within Section 11 (3) of the Insolvency Act 1986.
- Even if an adjudication itself is not a process giving rise to legal liability, what is the legal status if underlying conditions of contract or adjudication proceedings state (or a default situation arises where) the decision becomes final and binding, or agreement is reached by the Parties to that effect (although in the latter case if agreement is reached after discussion with Insurers there is no problem).

Insurers consider that they do not cover commercial decisions, only those arising out of applying the law to the relevant facts. The following are examples of concerns:

- Some adjudication procedures may specifically permit decisions using a fair and commercially reasonable view which may not coincide with there being a legal liability, so anyone using them has an immediate specific potential insurance cover problem.
- Despite no express provision for it within adjudication procedures, an adjudication decision may actually be made on commercial grounds (particularly likely to be the case with complex claims and the limited timetable), not solely on the facts and law (which is a requirement of most adjudication procedures). However, that might not be ascertainable from the decision, particularly if no reasons are given, and so Insurers in the absence of evidence to the contrary would have to assume their clients had a legal liability and make payment.
- There is a further concern that the consequence of an Insured commencing adjudication for, say, the payment of outstanding fees, may be a mid-adjudication cross-claim/statement allowed by an Adjudicator for incomplete or sub-standard work. This could lead to a 'loss' that the Insured might try to recover under a PI policy.

Insurers' reactions

Insurers' reactions have included to:

(a) Consider increasing excesses, reducing the top level of cover and increased premiums.
(b) Require immediate (e.g. 24 to 48 hours) notice to be provided to Insurers of an adjudication commencing (the period sometimes including, or at least not being very clear on whether it is including, weekends and/or public holidays).
(c) Instruct their own claims departments and advisors such as loss adjusters and experts as soon as possible (even to the extent of expending what might turn out to be wasted front-end costs if, say, the dispute is otherwise settled or a decision is in their favour), in order to determine when appropriate:
  (i) If a dispute could be covered by insurance (including whether a dispute value might be above the insured excess) and hence of

concern and something that Insurers should attempt to become involved in and influence.

(ii) If so, after the adjudication:
- Whether to appeal on an adverse decision, and on what grounds.
- Whether to try for a different decision in arbitration or litigation. This could be based on an assessment of whether the decision was complex and the defence not properly presented by the Insured or decided by the Adjudicator, the amount of money on balance being worth risking more money on, how unexpected the Adjudicator's decision was and so on, perhaps after the collection of any further data.

(d) Require Insureds to have the underlying construction contract state that disputes must be 'resolved by arbitration and that such arbitration is to be deferred until after practical completion.' This is known as a deferred arbitration clause, and permits Insurers to undertake further investigations before being obliged to hand over money (if that is what their policy would otherwise state was necessary).

(e) Agree a formal reservation of rights letter or equivalent policy amendment with the Insured if an Insured wants quick payment by Insurers, possibly before Insurers have had proper time to investigate whether cover is valid (e.g. the legal liability issue again) or whether there has been a breach of policy conditions (e.g. inadequate notice given).

(f) Not permit clauses in conditions of contract or adjudication procedures that make adjudication decisions 'final and binding' and which would prevent Insurers taking matters further, nor allow Parties to agree such settlements without the confirmation of Insurers, for the same reason. It is also possible that 'final and binding' decisions might make it more difficult for Insurers to recover money from others who might be liable to some extent.

(g) Consider whether the Insured's adjudication defence costs are covered (normally, but not always, limited to the Insured's own costs, not those of the other Party even if the Insured is the 'losing' Party). On a similar point, Insurers have to consider how the costs (including those of the Insured) of any subsequent appeal or arbitration/litigation proceedings will be apportioned.

Specimen Adjudication Endorsement – PI Insurance Policy

Adjudication Notifications

Any 'Notification of Adjudication' or a 'Referral Notice' pursuant to The Scheme For Construction Contracts (England and Wales) Regulations 1998 and/or The Scheme For Construction Contracts (Scotland) Regulations 1998, and/or any adjudication notice required by particular conditions of contract, must be notified within 24 hours in writing to any one of the Panel Solicitors listed on the attached schedule, or to any other firm of solicitors as may have been expressly agreed in writing between the Insured and the Insurers.

The foregoing adjudication notifications will be considered as notifications to Insurers and shall be subject to all other policy terms and conditions.

Adjudication-related information

All other circumstances, claim and material facts must be notified to Insurers in accordance with the terms and conditions of this PI Policy.

Final and binding decisions

It is a condition precedent to coverage being afforded that the Insured does not agree to accept the decision of the Adjudicator as finally determining the dispute without the prior consent in writing of the Insurers.

Insurers may pursue on Insured's behalf

It is agreed that the Insurers shall be entitled to pursue legal, arbitration or other proceedings in the name and on behalf of the Insured to challenge, appeal, open up or amend any decision, direction, award, or the exercise of any power of the Adjudicator, or stay the enforcement of any decision, direction, award or exercise of any power of the Adjudicator.

Assistance to Insurers

The Insured shall give all such assistance as the Insurers may reasonably require in relation to the foregoing proceedings.

Subrogation

For the avoidance of doubt this endorsement does not in any way limit the Insurers' right of subrogation.

Cover not extended

This endorsement does not extend the cover provided by this PI Policy.

Extent of Insurers' influence

Insurers can only influence their Insureds, and that is limited to what is agreed between them in the insurance policies. For example:

- Insurers may wish to be involved in the selection of specific adjudication procedures or clauses (where the Scheme does not apply), but:
  - Adjudication procedures are already pre-selected prior to insurance arrangements being finalised for standard conditions of contract.
  - Ad hoc adjudication procedures usually favour the Referring Party or Responding Party, depending on which side the drafting body feels it more likely to be on, but a Party must then suffer the consequences of such unequal weighting if it is on the 'wrong' side of a dispute.
- Insurers will generally, therefore, find that they will have little, if any, influence on the selection or wording of adjudication procedure clauses.
- Insurers or their legal advisors will usually have no choice on who is selected as the Adjudicator, except, perhaps, where a pre-contract nomination is made

- Insurers cannot interfere directly with an active adjudication process which is being controlled by the Adjudicator in accordance with the conditions of contract and/or adjudication procedures, e.g. an Adjudicator will have no reason or obligation to meet Insurers or their legal advisors, or to hear their views.
- The best that Insurers can do is become part of a Party's background advisory team, perhaps becoming involved in strategy such as the extent of non-cooperation with the Adjudicator, appealing, further investigations and so on, but that is of no additional immediate concern to the Adjudicator.

Concerns of Insureds

Some concerns for Insureds should include:

- Consideration of the preceding points in relation to the Insureds' own policies, e.g. where Insurers have onerous adjudication cover requirements, to fully understand what these mean.
- That a person at any level receiving a document which includes a notification of an intention to adjudicate (which may not always be clear) understands that Insurers may need to be notified, probably on a very urgent basis, and then takes, or passes it to someone else to take, appropriate action.
- A need to have specific persons nominated and procedures in place to rapidly confirm whether an adjudication dispute will involve an insurance claim and then to comply with any specific adjudication-related requirements of the policy, probably in the first instance relating to rapid notification provisions. This will include knowing who to notify, e.g. at Brokers or Insurers and having fall-back contact staff if primary contacts are not available.
- If there is no immediate indemnity included in their insurance policy, how to fund the amount of an Adjudicator's decision without the use of insurance money, perhaps until issues have been resolved finally by arbitration or the courts later in a project.

Chapter 17

Conclusions

Some contracts will not fall within the Construction Act, in which case, whatever adjudication provisions are in such contracts must apply. Where they do come under the Construction Act, the Act states minimum adjudication requirements that must be met. If they do not comply with the requirements then the Parties have an entitlement to refer a dispute to a statutory Scheme adjudication procedure. If contractual adjudication provisions comply with those minimum requirements then their provisions, even where significantly enhanced, must be followed.

There can thus be a theoretically unlimited number of possible contractual adjudication procedures instead of a more readily understood single statutory adjudication procedure. These might be entirely 'bespoke', might involve standard conditions of contract such as those of the ICE, JCT and GC/Works or they might involve the incorporation of adjudication procedures such as those produced by the CIC, as described within this text.

Further, adjudication procedures might involve by default or incorporation a version of the statutory Scheme itself, there now being Schemes for England and Wales, for Scotland and for Northern Ireland. Also, each Parliament or Assembly can make amendments to its own Scheme.

Additional complications are arising from court decisions. For example, some decisions relate only to contractual or only to statutory adjudications. Even within those two areas there may be regional differences, for example, between the English and Scottish courts, as described in the text.

Whatever its precise legal status, statutory adjudication was initially greeted by some as an unwarranted interference with the British principle of freedom of contract and an early demise was predicted. However, adjudication, was chosen because it was a pre-existing concept, in fact, one which was already incorporated into the New Engineering Contract. (It is, perhaps, ironic that the NEC was then the first standard form of contract that was decided by the courts to fall foul of the minimum requirements of the Construction Act!)

Construction adjudication has, in fact, proved to be a success, not as a result solely of legislation, but also from the courts imposing on Adjudicators a requirement to proceed in a way that meets at least the minimum common law standards of natural justice. As a consequence, adjudication avoids the slowness, formality and cost of litigation and arbitration, arriving at impartial and fair decisions, the majority of which are being accepted as a final determination by both Parties.

Appendix 1

Example agreement

To: (Name and address of Referring Party)

. . . .

(Name and address of Responding Party)

. . . .

Date:

Dear Sirs

[CONTRACT NAME . . . .]

ADJUDICATION BETWEEN . . . . (REFERRING PARTY)

AND . . . . (RESPONDING PARTY)

RELATING TO DISPUTE OVER . . . .

ADJUDICATOR'S AGREEMENT

1. On . . . . in accordance with the selection procedure laid down in . . . . , I was requested by . . . . to act as Adjudicator on the above dispute. I accept the appointment.

2. I shall conduct the dispute in accordance with the . . . . adjudication procedures as stated in [Relevant conditions of contract].

3. The attached terms and conditions shall apply to my appointment. They shall take precedence over any corresponding terms and conditions in the conditions of contract and adjudication procedures/rules pertaining to the conditions of contract.

4. The address and contact details for all future communications to me shall be:

. . . .

. . . .

Yours faithfully

(Adjudicator)

TERMS AND CONDITIONS OF APPOINTMENT

[The Adjudicator is to select whichever of the following is required]

1. The Adjudicator shall be paid a fee rate of [£100.00] per hour, excluding/including VAT. Time spent travelling shall be paid at the same rate.
2. The Adjudicator may call upon the assistance of other [Name of organisation] in-house administrative and technical staff from time to time for document and evidence handling, checking, analysis etc., with a view to processing such tasks more cheaply and quickly. Those staff will be under the control of the Adjudicator and the Adjudicator shall remain fully responsible for their output. Fee rates shall be determined in accordance with individual experience, qualifications and standing in [Name of organisation].
3. The Adjudicator shall be reimbursed all reasonable expenses including those for printing, reproduction or the purchase of documents; faxes and telephone calls; postage and delivery charges; travelling, meals and hotel costs; charges for obtaining legal or technical advice, and the costs (including fees and expenses) of any in-house support staff.
4. The Adjudicator is/is not currently registered for VAT. If appropriate, VAT shall be charged at the rate current on the date of invoice.
5. The Adjudicator and Parties shall keep the adjudication confidential, except as far as it is necessary to implement or enforce the Adjudicator's decision.
6. The Adjudicator shall be paid an advance Appointment Fee of [£5,000.00]. This Fee shall become payable in equal amounts by each Party within [7] days of the date of acceptance by the Adjudicator of the Adjudicator appointment. It will be deducted from the final amount of fees and expenses due. If the final amount due is less than the Appointment Fee the balance shall be refunded to the Parties.

   Any Party may pay the full Appointment Fee in order for the adjudication to proceed and this will be taken into account by the Adjudicator in the calculation of the amounts of Adjudicator's full fees and expenses to be paid by each Party.
7. All payments, other than the Appointment Fee, shall become due [10] days from the date of invoice, although earlier payment will be accepted, particularly if the prompt release of the Adjudicator's decision is required in accordance with Paragraph 12.

   Interest shall be payable at [8%] per annum above the Bank of England base rate for every day beyond the due date that the amount remains outstanding.

   Where an agreement is reached that the Adjudicator may have longer than 28 days to reach a decision, the Adjudicator may invoice on an interim monthly basis in addition to invoicing once the decision has been reached.
8. The Adjudicator shall act impartially and shall exercise reasonable skill and care in the performance of its duties.
9. The Parties may at any time agree to revoke the appointment of the Adjudicator.
10. The Adjudicator may resign where:

(i) In the Adjudicator's opinion the dispute varies significantly from that originally referred to the Adjudicator for confirmation of willingness to act and the Adjudicator at its sole discretion now considers that it is not now competent or able to decide it.
(ii) The full Appointment Fee has not been paid within the prescribed time.
(iii) The Adjudicator becomes aware of interest, financial, or otherwise in a matter relating to the dispute which the Adjudicator considers may cast reasonable doubt on the impartiality of the Adjudicator's decision.
(iv) The combining of two or more disputes leads to effective revocation of the Adjudicator's appointment due to another Adjudicator now dealing with the dispute.
(v) The Adjudicator becomes aware after appointment that the Referring Party has not followed correctly all pre-dispute contract procedures either in terms of establishing that an adjudicable dispute as defined in the contract exists, or in the selection of the Adjudicator.
(vi) Payment of any interim invoice issued in accordance with Paragraph 7 has not been received within [10] days from the date of issue.
(vii) After appointment one or both of the Parties wants for any reason to extend the timetable for the decision in accordance with the Contract and the Adjudicator, at its sole discretion, considers that it is unable to comply with those amended timetable requirements.
(viii) An Agreement is of a form that requires signature by the Parties, and the signatures are not forthcoming within [14] days of the date of the Adjudicator's appointment, or there is non-compliance with any alternative stated by the Adjudicator.

11. The Adjudicator must resign where it becomes apparent that the dispute is the same or substantially the same as one previously decided upon by an earlier Adjudicator and that, in the Adjudicator's view, the adjudication procedure precludes dealing with it [e.g. adjudication procedures such as the Scheme, Paragraph 9 (2)].
12. The Adjudicator's full fees and expenses, including any amounts outstanding from interim payment invoices issued in accordance with Paragraph 7 and including the costs of any technical and /or legal advice and any in-house assistance in accordance with Paragraph 3, shall be paid before the Adjudicator's decision is released.
13. Other than where either or both Parties obtain a legal decision that the Adjudicator has acted in bad faith, the Adjudicator shall be entitled to reimbursement of full fees and expenses incurred up until the date of receipt by the Adjudicator from the Referring Party of notification of any settlement between the Parties, or knowledge of revocation or resignation in accordance with Paragraphs 9, 10 and 11, or release of the decision in accordance with Paragraph 12. In addition, the Adjudicator shall be entitled to reimbursement for any reasonable fees and expenses incurred for preparing a final account, returning the balance of any advance, returning or destroying documents as instructed and similar minor winding-up activities.

14. The Parties shall be jointly and severally liable for the payment of outstanding Adjudicator's fees and expenses. The Parties shall pay the fees and expenses in equal shares (unless the Adjudicator in the Adjudicator's decision directs otherwise), subject to any adjustments that need to be made where a Party or Parties have not paid their share of any advance Appointment Fee. Any Party or Parties may, however, pay outstanding fees and expenses and seek to recover any appropriate contribution from other non-paying Party or Parties as a debt due.
15. The Adjudicator shall not be liable for anything done or omitted in the discharge or purported discharge of its functions as Adjudicator unless the act or omission is in bad faith, and any employee or agent of the Adjudicator shall be similarly protected from liability. The Parties shall save harmless and indemnify the Adjudicator and any employee or agent of the Adjudicator against all claims by third parties, including negligence, and in respect of this indemnity shall be jointly and severally liable. For the avoidance of doubt in relation to the Construction (Rights of Third Parties) Act 1999, the Parties to this Adjudicator Agreement do not intend any of its terms to be enforceable against the Adjudicator by any third party.
16. The Adjudicator shall be advised by the Parties within four weeks of completing its decision whether either Party wishes to have returned to it any of the documents provided by that Party to the Adjudicator. In the absence of such advice the Adjudicator may destroy the documents. Where a Party wishes the Adjudicator to retain any of its documents for a longer period, that shall be at the discretion of the Adjudicator.

Appendix 2

Example Adjudicator's first letter

[No two adjudication procedures are identical. The following is based on the ICE Adjudication Procedure which is different from others, particularly in the matters that lead up to an adjudicable 'dispute' arising. It is intended only as an aide-mémoire for an Adjudicator of matters it might want to include and then be adapted for non-ICE procedures.]

To: (Name and address of Referring Party)

. . . .

(Name and address of Responding Party)

. . . .

Date:

Dear Sirs

[CONTRACT NAME . . . .]

ADJUDICATION BETWEEN . . . . (REFERRING PARTY)

AND . . . . (RESPONDING PARTY)

RELATING TO DISPUTE OVER . . . .

ADJUDICATOR'S COMMUNICATION NO. 1

1. A matter of dissatisfaction has been decided by the Engineer/has not been decided by the Engineer within the required timescale.

2. A Notice of Dispute dated. . . . . .has been served.

3. Before a Notice to Refer to Arbitration has been served the . . . . Party asked the . . . . Party if the dispute could be dealt with by conciliation

in accordance with the current version of the ICE Conciliation Procedure. The............ Party agreed and the Conciliator's recommendation was received by the [current dissenting] Party on .... Clause 66 (5)(b) of the ICE Conditions of Contract provides that the recommendation of the Conciliator finally determines the dispute if the requisite notice to refer the dispute to adjudication or to arbitration is not served within one month of the recommendation.

4. A written Notice of Adjudication dated .... has been given by the Referring Party to the Responding Party relating to a dispute over

   ....

   This is within the one month timescale described in the preceding paragraph.

5(a) I have, under paragraph 3.1 of the ICE Procedure, been named in the Contract as an/the Adjudicator.

Or:

I have, under Paragraph 3.1 of the ICE Procedure, been agreed between the Parties prior to the issue of the Notice of Adjudication as an/the Adjudicator.

5(b) A copy of the Notice of Adjudication was received by me on .... requesting confirmation within four days of the date of issue that I was able and willing to act.

On................I confirmed that I was.

Or:

5(c) I confirm that, under Paragraph 3.2 of the ICE Procedure, I have been named in the Notice of Adjudication as a person acceptable to the Referring Party who has agreed to act as Adjudicator. The Responding Party, in writing to me at the address given in the Notice of Adjudication on ...., has selected me as the Adjudicator. [*This notification should be within four days of the date of issue of the Notice of Adjudication. What happens if it is not, even by one or two days, is discussed in the main text.*]

Or:

5(d) It is understood that after four days of the date of the Notice of Adjudication no confirmation was received under Paragraph 3.1 of the ICE Procedure, or no selection was made under Paragraph 3.2, or the named Adjudicator did not accept or was unable to act. I have seen documentation which shows that, as a consequence, the Referring/Responding Party on .... [*Must be within a further three days*] requested [(a) the appointing body named in the Contract, the .... (b) The Institution of Civil Engineers,] to appoint an Adjudicator. By copy letter from the [(a) appointing authority (b) The Institution of Civil Engineers] dated .... I note that the Referring and Responding Parties were informed of my appointment.

6. I acknowledge receipt from the Referring Party on .... of the whole or last of the following Referral Documents:

   (a) A copy of the Notice of Adjudication including:

   - Details and date of the Construction Contract between the Parties
   - The issue(s) the Referring Party requires the Adjudicator to decide
   - Details of the nature and extent of redress being sought.

   (b) A copy of the adjudication provisions contained within the construction contract.

   (c) The information upon which the Referring Party relies, including supporting documents.

[Although neither the ICE Procedure Paragraph 2.1, nor Paragraph 4.1 (from which the foregoing list was derived) state so, it seems important that the Adjudicator is also provided specifically with enough evidence of (a) any Conciliator recommendation, (b) the Engineer's decision, (c) that the decision (and any recommendation) is disagreed with by the Referring Party, (d) that any critical timetable requirements pertaining to each have been complied with, and (e) the issue was one arising under the contract, to be able to confirm that an adjudication ICE 'dispute' has, in fact, arisen. If this information is not provided (although not stated in Paragraphs 2.1 or 4.1), then it might be claimed that the Date of Referral does not commence until it is.

It is also seems important that the pre-appointment documentation leading to the appointment under one of the para 5 options of this letter are also supplied in order for the Adjudicator to confirm to its own satisfaction that the appointment process was properly carried out and that there are no problems over jurisdiction from that point of view.]

7. These Referral Documents are taken as a full statement of the Referring Party's case and the aforementioned date of their receipt by me is the Date of Referral for the purposes of this Adjudication.

8. The Referring Party should have copied the Referral Documents to the Responding Party. Should the Responding Party not have received a full set of copies of the Referral Documents by the same date as myself, any subsequent delay in their receipt and the consequences as notified to me may be taken into account (where considered appropriate by me) in my apportionment of my fees and expenses between the Parties.

9. The names and addresses at the top of this letter shall be assumed to be the ones to be used by me and the Parties for the service of notices and documents, until otherwise notified by either of the Parties. My corresponding address is given on the letterhead of this letter. Send your notices and documents clearly marked as 'Confidential' for the attention of [Name of Adjudicator]. As the dispute is confidential, no further identifying description should be provided on any outer cover.

10. Based on the information provided, I confirm that I am not an employee of either of the Parties nor do I knowingly have any interest, financial, or otherwise, in any matter relating to the dispute.

[Paragraph 108 (2)(e) of the Construction Act requires an Adjudicator to act impartially. An interpretation of this is given in paragraph 4 of the Scheme, which requires an Adjudicator not to be an employee of any of the Parties and to 'declare any interest, financial or otherwise in any matter relating to the dispute'. The ICE Procedure Paragraph 1.3 also requires the Adjudicator to act impartially. To reduce the possibility of a doubt being raised concerning Adjudicator impartiality, meeting at least the declaration requirements of the Scheme appears sensible practice.]

11. The Referring Party does not wish to exercise its statutory right to refer the dispute, which arises under the contract, for adjudication under a procedure complying with Section 108 of the Housing Grants, Construction and Regeneration Act 1996 (the Act). Instead, the Referring Party wishes to use the adjudication provisions of the contract, i.e. those incorporating the ICE Adjudication Procedure, 1997.

    I have studied appropriate parts of the Referral Documents and note:

    - An Engineer's decision on a matter of dissatisfaction has been given which is not accepted by one of the Parties/has not been provided in the required timescale.
    - A Notice of Dispute has been served.
    - A Notice of Adjudication has been served.
    - The matter of dissatisfaction concerns an issue which arose *under* the Contract, as required by paragraph 2.1 of the ICE Procedure. [*This may not be important – see comments in Chapter 3 on the ICE Procedure, Clause 66 (6)(a), Paragraphs 1.2, 2.1, 5.2, and 5.3, as well as Chapter 2 on the Construction Act , Section 108 (1).*]
    - A dispute has therefore arisen which can be dealt with under the adjudication provisions of this Contract.

    [*For other circumstances it might be appropriate to add or substitute one or more of the following:*

    - *The Construction Contract is a 'construction contract' in accordance with Sections 104 to 106 of the Act*

- *The Construction Contract is in writing in accordance with Section 107 of the Act*
- *The Construction Contract does not appear to comply with minimum requirements of Section 108 of the Act because ....*
- *The Referring Party wishes to exercise its right to adjudication under the Act and requires the dispute to be resolved using The Scheme for Construction Contracts (England and Wales) Regulations 1998 (the 'Scheme') and any subsequent amendments to it.]*

12. To the best of my knowledge, information and belief, an Adjudicator has not previously provided a decision on the dispute matter and so my dealing with the dispute is not an issue that requires the agreement of both Parties in accordance with Paragraph 5.3 of the ICE Procedure. Either of the Parties should advise me immediately if they consider that not to be the case.

[Paragraph 5.3 of the ICE Procedure permits the Adjudicator to readdress a previous adjudication decision but only if agreed by the Parties. This can include a decision on an interim issue that might alter over the duration of the Contract (e.g. a final extension of time determination). As discussed in the main text, there remains a problem if a decision has been agreed by the Parties to be final and binding in accordance with Paragraph 1.5 or by default in accordance with the ICE Conditions of Contract Clause 66(9)(b), although under Paragraph 1.1 it states that the ICE Procedure takes precedence over the contract Clauses and hence Paragraph 5.3 would apply. An Adjudicator might, however, consider it prudent not to become involved in any issue that is stated to be final and binding, or conclusive until this issue is better clarified.]

13. To the best of my knowledge, information and belief, the disputed matter is as previously notified in the Notice of Dispute/Notice of Adjudication so can be dealt with under the ICE Adjudication Provisions.

[Paragraph 5.2 permits other matters to be within the scope of adjudication if all Parties and the Adjudicator agree.]

14. To the best of my knowledge, information and belief, my appointment as Adjudicator followed the procedures laid down by the Contract.

[If it does not, then some words are necessary to explain why the Adjudicator is proceeding, e.g. 'there was a slight delay in the proposed timetable but I am continuing in the spirit and intention of Clause xxxx of the Act in order to minimise costs and provide a quick decision'.]

15. I am therefore proceeding as the Adjudicator for this dispute.

16. I will provide Adjudicator's Instructions when I consider that these are necessary. My first instructions are given as an Attachment to this letter.

[*Check this is applicable.*]

17. Paragraph 1.5 of the ICE Procedure is drawn to your attention. This states that the Parties can agree that the Adjudicator's decision will be a final determination of the dispute. If this is not already agreed within the Construction Contract or is (or is not but can be) formally recorded between you elsewhere, and it accords with your wishes, I will be agreeable to receiving copies of your written agreement and incorporating reference to it within my decision.

18. Clause 66 (9)(b) of ICE Conditions of Contract is also drawn to your attention. It states that if a Notice to Refer to arbitration is not served within three months of receipt of an Adjudicator's decision, then that decision shall become final and binding.

Yours faithfully

(Adjudicator)

Appendix 3

Example Adjudicator's first instructions

[This is suggested as the basis of a possible attachment to Adjudicator's Communication No. 1 but may be incorporated into the main text of that Communication or be written as a separate letter. It will be apparent that for many disputes the Instructions can be simplified. The items given are therefore only supplied as an aide-mémoire which includes issues that might need considering in more complex disputes.]

[CONTRACT NAME . . . .]

ADJUDICATION BETWEEN . . . . (REFERRING PARTY)

AND . . . . RESPONDING PARTY)

RELATING TO DISPUTE OVER . . . .

FIRST ADJUDICATION INSTRUCTIONS

My First Instructions are that:

(a) The Referring Party shall provide a Summary Referral Statement of no more than 2,000 words outlining and cross-referenced to the case made in the Referral Documents, for my receipt within 7 days of the Date of Referral.

(b) The Responding Party shall provide any response, the Response Documents, to the issue(s) raised by the Referring Party in the Referral Documents, for my receipt within [14] days of the Date of Referral. [*14 days is a specific requirement of Paragraph 5.4 of the ICE Procedure but it might vary for other adjudication procedures, e.g. the Scheme does not mention a response at all.*] This shall be accompanied by a Summary Response Statement of no more than 2,000 words outlining and cross-referenced to the Response Documents. Issues or amounts included within the Referral or Response Documents, not responded to, shall be considered to be agreed.

(c1) [Replies by the Referring Party to the response may need to be addressed.]

(c2) *[How to deal with any counter-claims in the response may also need to be considered.]*

(d) Should I consider oral evidence or presentations are required, a Party shall not be represented by more than one person at any one time unless I otherwise instruct.
[*This is as Paragraph 16(2) of the Scheme, so it might be difficult to argue that it is particularly unreasonable.*]

(e) Legal issues that may arise during the adjudication which are related to the dispute, but not to the interpretation of the Act or adjudication procedures, must be provided in writing to me and the other Party (or Parties). No oral presentation of legal arguments shall be permitted.

(f) Unless expressly requested by the Parties for me to decide, legal issues and points related to the interpretation of the Act or adjudication procedures in force shall be matters for the Parties, not for this adjudication.

(g) I shall proceed with the adjudication on the basis that the Referring Party is satisfied that there is a bona fide 'dispute' and that its decision to issue a Notice of Adjudication and subsequent adjudication procedures, including my appointment, are correct.

(h) All documents sent to me from the date of this instruction shall also be copied in full (including all attachments) to the other Party (or Parties) so that they are recorded as received on the same date that I receive them. If there is any late delivery and if a Party considers there has been a disadvantage incurred, full details with appropriate evidence including a record of delivery shall be supplied to me and copied to the other Party (or Parties) as soon as reasonably possible. If considered valid by me, and where not excluded by the adjudication procedures, I may take account of the disadvantage in the apportionment of my fees and expenses between the Parties.

(i) Before releasing my decision, the Parties shall provide full payment of my fees and expenses (including the costs of obtaining any legal and/or technical advice). Any Party may pay these fees and expenses in order to obtain my decision.

(j) Enclosed with this letter is a copy of my Adjudicator's Agreement and my associated Terms and Conditions. The Parties must sign their copy of the Agreement and return it to me with the Appointment Fee within [7] days of the date of this letter.

The Terms and Conditions contain details of an advance Appointment Fee which will be deducted from the amount of final fees and expenses due. The Parties shall pay their proportion of that Fee within [7] days of the date of my appointment. For ease of reference, the appointment date is.............. Any Party may pay the full Appointment Fee and that will be taken account of by me in my calculation of the fees and expenses due from each Party. Failure to pay the Appointment Fee within the required timescale will, at my discretion, constitute a revocation of my appointment in accordance with Paragraph 10 of the Terms and Conditions and I shall be entitled to be paid fees and expenses incurred to that date.

(k) I may provide further instructions in due course relating to how this adjudication shall proceed.

(Adjudicator) Date:

Appendix 4

Example advice to lay clients

[Possible attachment to Adjudicator's Communication No. 1. It is based primarily on the ICE Adjudication Procedure. It can be adapted for any other procedures.]

[CONTRACT NAME...]

ADJUDICATION BETWEEN .... (REFERRING PARTY)

AND .... (RESPONDING PARTY)

RELATING TO DISPUTE OVER.......................................

GUIDANCE ON SOME PROCEDURES THAT WILL BE FOLLOWED IN THIS ADJUDICATION

For information and guidance on the procedures that I presently intend that I and the Parties shall follow, I provide some extracts, occasionally paraphrased, from the [ICE Adjudication Procedure], with some comments and/or interpretations italised in square brackets. The Parties should obtain their own full copy of the [ICE Adjudication Procedure] and advice on interpretation.

- The Adjudicator shall have complete discretion as to how to conduct the Adjudication, and shall establish the procedure and timetable, subject to any limitation that there may be in the construction contract or Part II of the Housing Grants, Construction and Regeneration Act 1996 [the Construction Act]. The Adjudicator shall not be required to observe any rule of evidence, procedure or otherwise, of any court. Without prejudice to the generality of these powers, the Adjudicator may [*or may not*]:

  (a) Ask for further written information.
  (b) Meet and question the Parties (and any representatives).
  (c) Visit the site.
  (d) Request the production of documents or the attendance of people whom the Adjudicator considers could assist.
  (e) Set times for (a) – (d) and similar activities.

(f) Proceed with the adjudication and reach a decision even if a Party fails:
  (i) to provide information;
  (ii) to attend a meeting;
  (iii) to take any other action requested by the Adjudicator.
(g) Issue such further directions as the Adjudicator considers to be appropriate.

[There is a limited timescale available for adjudication and a requirement for it to provide a relatively quick and cheap decision. I will therefore allow both Parties to present their cases, and may permit them to respond to each other's cases at my discretion taking the circumstances, complexity, timetable of the dispute, my availability and so on into account. Parties may be met and interviewed separately, but I will communicate any information gained that might be used in reaching my decision to the other Party or Parties. Similarly, any additional legal or technical advice obtained in accordance with the pertaining adjudication procedure will be communicated to all Parties before my producing a decision.]

- The Adjudicator may obtain legal or technical advice having first notified the Parties of his intention. *[i.e. I do not need the Parties' agreement on this.]*

- The Adjudicator may take the initiative in ascertaining the facts and the law, and may rely on its own expert knowledge and experience. *[The initiative is optional and will depend on the complexity of the claim, the amount of information supplied by the Parties, how much I wish to rely on my own expertise and knowledge (however limited), the required timetable and so on.]*

- The Parties shall implement the Adjudicator's decision without delay whether or not the dispute may be later referred to arbitration or legal proceedings. Paragraph 6.7 entitles Parties to seek summary enforcement of the decision. *[Note that I will not be involved in enforcing compliance with my decision. Non-compliance is a contractual breach for which enforcement procedures exist in the courts, whether or not an entitlement to summary judgment has been provided.]*

- The Adjudicator must reach its decision within 28 days of the Date of Referral *[this is the date on which I received a copy of the Referring Party's full statement of case]* unless the Parties have agreed a longer period, or may be extended by 14 days with the consent of the Referring Party only. *[An extension is an option for me to accept or not.]*

- The Adjudicator may open up, review and revise any previous decision, opinion, instruction, direction, certificate or valuation made 'under or in connection with' the contract and which is relevant to the dispute.

- Unless otherwise agreed between the Adjudicator and both Parties, the Responding Party may submit a response to the Referring Party's case within 14 days of the Date of Referral. *[If you wish to respond, I confirm that I shall certainly require that response within [14] days.]*

- The Adjudicator is not required to give reasons for its decision.

- The Adjudicator's appointment shall be on the terms and conditions of a standard pro forma ICE Adjudicator's Agreement and associated Schedule. The Parties shall sign the Agreement within 7 days of being requested to do so.

- The Parties shall bear their own costs and expenses and shall be jointly and severally responsible for the Adjudicator's fees and expenses (which include those of any legal or technical advisor *[and that of any in-house assistance under my control and for which I will be responsible that I consider might expedite my involvement in the adjudication more quickly and economically]* appointed by the Adjudicator).

- At any time until 7 days before the Adjudicator is due to reach its decision, the Adjudicator may give notice to the Parties that it will deliver its decision only on full payment of its fees and expenses. This includes the cost of any legal and technical advice *[and any in-house administration and technical support]* obtained by the Adjudicator. Any Party may then pay these costs in order to obtain the decision and recover the other Party's share of the costs as a debt due. *[I advise now that I shall definitely require payment before I release my decision.]*

- *[All documents sent to me during the Adjudication must be copied to the other Party so that the Party and Adjudicator can, as far as reasonably possible, receive those documents on the same day. Delays in receipt of documents by me or a Party may lead to some delays in providing my decision or some associated fees being awarded against the Party sending the documents.]*

- *[To ensure that any allegations of delay can more easily be assessed …]* all notices shall be sent by recorded delivery. *[This does not only mean a formal service operated by the postal service but includes any means that results in a formal record of receipt being obtained by a Party.]*

Appendix 5

Example issues to be addressed in a decision

To: (Name and address of Referring Party)

. . . .

(Name and address of Responding Party)

. . . .

Date:

Dear Sirs

[CONTRACT NAME . . . .]

ADJUDICATION BETWEEN . . . . (REFERRING PARTY)

AND . . . . (RESPONDING PARTY)

RELATING TO DISPUTE OVER. . . . .

DECISION

[It is suggested that a decision should consider, albeit as briefly and succinctly as is reasonable (bearing in mind that the courts might refer to the document in enforcement proceedings), the following headings.]

A. Introduction

1. Underlying conditions of contract
2. Adjudication Procedures/Rules that apply
3. Disputes to be decided [usually from the Notice of Adjudication]
4. How the Adjudicator's appointment arose
5. Adjudicator's Agreement
6. Date referral commenced

7. Timetable for decision and any agreed amendments to it
8. Any agreement by Parties that a decision is to be final and conclusive

B. Information/evidence

9. Surrounding circumstances and events leading up to disputes
10. Documents received from Referring Party
11. Documents received from Responding Party
12. Meetings held
13. Visits to site undertaken
14. Outside legal/technical advice received
15. Points of law raised by Parties and/or considered by Adjudicator

C. The decision(s) reached

16. Decisions reached on disputed points
17. Interest
18. VAT
19. Reasons for decisions reached (if reasons are to be provided)

D. Adjudicator's fees and expenses

20. Adjudicator's fees and expenses (including those for outside advice (if any))
21. Decision on apportionment of Adjudicator's fees and expenses (with adjustments for any advance payment or for payments made by one Party only)
22. Reasons for apportionment of Adjudicator's fees and expenses (if reasons are to be provided)

E. Party costs (if any)

23. Parties' costs (if to be decided by Adjudicator)
24. Decisions reached on apportionment of those costs
25. Reasons for apportionment of Parties' costs (if reasons are to be provided)

F. Signature and date of decision

26. Name and signature of Adjudicator   Date of Decision

Appendix 6

Non-adjudication cases

The following is a list of non-adjudication legal cases referred to in the text, given in the order of their first mention.

1. *Potton Developments Ltd* v *Thompson* [1998] DLSSCS 98
2. *Monmouth CC* v *Costelloe & Kemple Ltd* (1965) 5 BLR 1
3. *Fillite* v *Aqua Lift* [1986] 45 BLR 27
4. *Produce Brokers* v *Olympic Oil and Coke* [1916] 2 KB 296
5. *Antonis P Lemos* [1985] AC 711
6. *Abdullah Fahem & Co.* v *Mareb Yemen Insurance Co. and Tomen (UK) Ltd* [1977] 2 Lloyd's Law Rep 738
7. *Al-Naimi* v *Islamic Press Agency Incorporated* [2000] 1 Lloyd's Law Rep 522
8. *Empressa Exportadora de Azucar* v *Industria Azucarera Nacional SA* [1983] 2 LLR 171
9. *Harbour Assurance Company (UK) Ltd* v *Kansa General International Assurance Company Ltd* [1993] 3 All ER 897.
10. Court of Appeal in *Walters* v *Whessoe* (1960) 6 BLR 23
11. House of Lords in *Smith* v *South Wales Switch Gear* (1977) 8 BLR 1
12. House of Lords in *Beaufort Developments (NI) Ltd* v *Gilbert Ash (NI) Ltd* [1999] 1 AC 266
13. Court of Appeal in *Northern Regional Health Authority* v *Derek Crouch Construction Ltd* (1984) QB 644
14. Court of Appeal, *Brian Andrews* v *John H Bradshaw and H Randell & Son Ltd* [1999]
15. *R* v *Gough* [1993] AC 646
16. *Locabail (UK) Ltd* v *Bayfield Properties Ltd* [2000] 1 All ER 65
17. *Laker Airways Inc* v *FLS Aerospace Ltd* [2001] 1 WLR 113
18. *R* v *Sussex Justices ex parte McCarthy* [1924] 1 KB 256
19. *Wiseman* v *Borneman* [1971] AC 297 HL
20. *Hayter* v *Nelson* [1990] 2 LR 265
21. *Halki Shipping Corporation* v *Sopex Oils Ltd* (1998) 1 WLR 726
22. *Enco* v *Zeus* (1991) 28 Con LR 25
23. *Cott UK Ltd* v *F E Barber Ltd* : QBD 14 January 1997 [BLISS CLMK]
24. *Miller Civil Engineering Ltd* v *National Rivers Authority*: QBD July 1997 [BLISS CMAK]
25. *Dixons* v *Murray-Oboynski*, OR Court [1997]
26. *Crown House Engineering Ltd* v *AMEC Projects Ltd* [1989]

27. *Tubeworkers* v *Tilbury* (1985) 30 BLR 67
28. *Gilbert Ash* v *Modern Engineering* [1974] AC 689

Appendix 7

Adjudication cases

The following is a chronological list of the adjudication cases known to the authors at the time of writing. Full details of many can be obtained from:

- The Government Court Services website at: http://www.courtservice.gov.uk
- The Scottish website at: http://www.scotcourts.gov.uk
- Unreported cases may be found at: http://www.adjudication.co.uk

Where (E) and (S) are used, this indicates that the case concerned was heard in the English and Scottish courts respectively.

1. *Cameron (A) Ltd* v *John Mowlem*, 1990 52 BLR 24
2. *Drake & Scull Engineering Ltd* v *McLaughlin & Harvey plc*, 1992 60 BLR 102
3. *Mercury* v *Director General of Telecommunications*, 1994 1388 SJLB 183
4. *Macob Civil Engineering Ltd* v *Morrison Construction Ltd*, 12/2/99, BLR 93 (TCC)(E)
5. *Rentokil Ailsa Environmental Ltd* v *Eastend Civil Engineering Ltd*, 12/3/99, CILL 1999 1506 (S)
6. *Outwing Construction Ltd* v *H Randell & Son Ltd*, 15/3/99 BLR 156 (TCC)(E)
7. *Rentokil Ailsa Environmental Ltd* v *Eastend Civil Engineering Ltd*, 31/3/99, CILL 1999 1506 (S)
8. *A Straume (UK) Ltd* v *Bradlor Developments Ltd*, 7/4/99, CILL 1999 1520 (Chancery Division)(E)
9. *A & D Maintenance and Construction Ltd* v *Pagehurst Construction Services Ltd*, 23/6/99, CILL 1999 1518 (TCC)(E)
10. *Allied London & Scottish Properties plc* v *Riverbrae Construction Ltd*, 12/7/99, CILL 1999 1541, BLR 346 (S)
11. *The Project Consultancy Group* v *The Trustees of the Gray Trust*, 16/7/99, CILL 1999 1531, BLR 377 (TCC)(E)
12. *John Cothliff Ltd* v *Allen Build (North West) Ltd*, 29/7/99, CILL 1999 1530, BLR 426 (Liverpool County Court)
13. *Palmers Ltd* v *ABB Power Construction Ltd*, 6/8/99, CILL 1999 1543, BLR 426 (TCC)(E)
14. *Lathom Construction Ltd* v *Anne & Brian Cross*, 29/10/99, CILL 1999 1568 (TCC)(E)
15. *Homer Burgess Ltd* v *Chirex (Annan) Ltd (1)*, 10/11/99, CILL 2000 1580 (S)

16. *Bouygues (UK) Ltd* v *Dahl-Jensen (UK) Ltd* (1), 17/11/99, BLR 522, CILL 1999 1566 (TCC)(E)
17. *Homer Burgess Ltd* v *Chirex (Annan) Ltd (2)*, 18/11/99 (S)
18. *Sherwood & Casson Ltd* v *MacKenzie Engineering Ltd*, 30/11/99, CILL 2000 1577 (TCC)(E)
19. *Fastrack Contractors Ltd* v *Morrison Construction Ltd and Imregio UK Ltd*, 4/1/2000, CILL 2000 1589, BLR 168 (TCC)(E)
20. *VHE Construction plc* v *RBSTB Trust Co Ltd*, 13/1/2000, CILL 2000 1592, BLR 187 (TCC)(E)
21. *Northern Developments (Cumbria) Ltd* v *J & J Nichol*, 24/1/2000, CILL 2000 1601, BLR 158, (TCC)(E)
22. *Samuel Thomas Construction Ltd* v *Bick & Bick Ltd (t/a J & B Developments)*, 28/1/2000 (High Court Exeter)
23. *F W Cook Ltd* v *Shimizu (UK) Ltd*, 4/2/2000 (TCC)(E)
24. *Workplace Technologies plc* v *E Squared Ltd and Anor*, 16/2/2000, CILL 2000 1607 (TCC)(E)
25. *Atlas Ceiling & Partition Company Ltd* v *Crowngate Estates (Cheltenham) Ltd*, 18/2/2000 (TCC)(E)
26. *Nolan Davis Ltd* v *Steven Catton*, 22/2/2000, unreported (TCC)(E)
27. *Grovedeck Ltd* v *Capital Demolition Ltd*, 24/2/2000, CILL 2000 1604, BLR 181 (TCC)(E)
28. *Absolute Rentals Ltd* v *Gencor Enterprises Ltd*, 28/2/2000, CILL 2000 1687 (E)
29. *Bloor Construction (UK) Ltd* v *Bowmer & Kirkland (London) Ltd*, 5/4/2000, CILL 2000 1626, BLR 764 (TCC)(E)
30. *Bridgeway Construction Ltd* v *Tolent Construction*, 11/4/2000, unreported (TCC Liverpool)
31. *Tim Butler Contractors Ltd* v *Merewood Homes Ltd*, 12/4/2000, unreported (TCC Salford)(E)
32. *Edmund Nuttall Ltd* v *Sevenoaks District Council*, 14/4/2000, unreported (E)
33. *Herschel Engineering Ltd* v *Breen Property Ltd* (1), 14/4/2000 (TCC)(E)
34. *Strathmore Building Services Ltd* v *Colin Scott Greig (t/a Hestia Fireside Design)*, 18/5/2000 (S)
35. *John Mowlem & Company plc* v *Hydra-Tight Ltd (t/a Hevilifts)*, 6/6/2000, CILL 2000 1650 (TCC)(E)
36. *R G Carter Ltd* v *Edmund Nuttall Ltd*, 21/6/2000 (E)
37. *Stiell Ltd* v *Riema Control Systems Ltd*, 23/6/2000 (S)
38. *Christiani & Neilsen Ltd* v *The Lowry Centre Development Company Ltd*, 29/6/2000, unreported (TCC)(E)
39. *Nottingham Community Housing Association Ltd* v *Powerminster Ltd*, 30/6/2000 (TCC)(E)
40. *KNS Industrial Services (Birmingham) Ltd* v *Sindall Ltd*, 17/7/2000, 75 Con LR 71, (TCC)(E)
41. *Ken Griffin and John Tomlinson (t/a K & D Contractors)* v *Midas Homes Ltd*, 21/7/2000, unreported (E)
42. *Shepherd Construction Ltd* v *Mecright Ltd*, 27/7/2000, BLR 489 (E)
43. *Mitsui Babcock Energy Services Ltd*, Petitioner (Judicial Review), 27/7/2000 (S)

44. *Herschel Engineering Ltd* v *Breen Properties Ltd* (2), 28/7/2000, unreported (TCC)(E)
45. *Bouygues (UK) Ltd* v *Dahl-Jensen (UK) Ltd (2)*, 31/7/2000, CILL 2000 1616, 1BLR 49 (Court of Appeal)(E)
46. *ABB Power Construction Ltd* v *Norwest Holst Engineering Ltd*, 1/8/2000 (TCC)(E)
47. *George Parke* v *The Fenton Gretton Partnership*, 2/8/2000, CILL 1712, (E)
48. *Whiteways Contractors (Sussex) Ltd* v *Impresa Castelli Construction UK Ltd*, 9/8/2000 (TCC)(E)
49. *Stubbs Rich Architect* v *W H Tolley & Son Ltd*, 8/8/2001 (E)
50. *Discain Project Services Ltd* v *Opecprime Development Ltd (1)*, 9/8/2000, 8 BLR 402 (TCC)(E)
51. *Universal Music Operations Ltd* v *Flairnote Ltd and Anr*, 24/8/2000, unreported (E)
52. *Elenay Contracts Ltd* v *The Vestry*, 30/8/2000, CILL 2000 1679 (TCC)(E)
53. *Cygnet Healthcare plc* v *Higgins City Ltd*, 6/9/2000 (E)
54. *Woods Hardwick Ltd* v *Chiltern Air Conditioning*, 2/10/2000 (TCC)(E)
55. *Maymac Environmental Services Ltd* v *Faraday Building Services Ltd*, 16/10/2000, CILL 2000 1686 (TCC)(E)
56. *ABB Zantingh Ltd* v *Zendal Building Services Ltd*, 12/12/2000 (E)
57. *Karl Construction (Scotland) Ltd* v *Sweeney Civil Engineering (Scotland) Ltd*, 21/12/2000 (S)
58. *LPL Electrical Services Ltd* v *Kershaw Mechanical Services Ltd*, 2/2/2001 (E)
59. *Holt Insulation Ltd* v *Colt International Ltd*, 2/2/2001 (Liverpool TCC)
60. *Glencot Development and Design Co Ltd* v *Ben Barrett & Son (Contractors) Ltd*, 2/2/2001 and 13/2/2001 (E)
61. *Rainford House Ltd (In Admin Receivership)* v *Cadogan Ltd*, 13/2/2001, All ER (Digest) 144 (E)
62. *Staveley Industries plc (t/a EI.WHS)* v *Odebrecht Oil & Gas Services Ltd*, 28/2/2001 (E)
63. *Joseph Finney plc* v *Gordon Vickers and Gary Vickers (t/a The Mill Hotel)*, 7/3/2001 (E)
64. *Watson Building Services Ltd*, Petitioner (Judicial Review), 13/3/2001 (S)
65. *Austin Hill Building* v *Buckland Securities Ltd*, 11/4/2001 (E)
66. *Discain Project Services Ltd* v *Opecprime Development Ltd (2)*, 11/4/2001 (E)
67. *Farebrother Building Services Ltd* v *Frogmore Investments Ltd*, 20/4/2001 (E)
68. *RJT Consulting Engineers Ltd* v *D M Engineering (NI) Ltd*, 9/5/2001 (Liverpool TCC)
69. *Re A Company (no. 1299 of 2001)*, CILL 1745, 15/5/2001 (E)
70. *Faithful & Gould Ltd* v *Arcal Ltd*, 25/5/2001 (E)
71. *Fence Gate Ltd* v *James R Knowles Ltd*, 31/5/2001 (E)
72. *Barr Ltd* v *Law Mining Ltd*, 15/6/2001 (S)
73. *Sindall Ltd* v *Solland and Ors*, 15/6/2001 (E)
74. *C & B Scene Concept Design Ltd* v *Isobars Ltd (1)*, 20/6/2001 (E)
75. *Ballast plc* v *The Burrell Company (Construction Management) Ltd*, 21/6/2001 (S)
76. *William Naylor (t/a Powerfloated Concrete Floors)* v *Greenacres Curling Ltd*, 26/6/2001 (S)

77. *British Waterways Board*, Petitioner (Judicial Review), 5/7/2001 (S)
78. *City Inn Ltd* v *Shepherd Construction Ltd*, 17/7/2001 (S)
79. *Gibson Lea Retail Interiors Ltd* v *Macro Self-Service Wholesalers Ltd*, 24/7/2001 (E)
80. *David McLean Housing Contrs. Ltd* v *Swansea Housing Association Ltd*, 27/7/2001, (TCC)(E).
81. *SL Timber Systems Ltd* v *Carillion Construction Ltd*, 27/7/2001, CILL 1760 (S)
82. *Yarm Road Ltd* v *Costain Ltd*, 30/7/2001 (E)
83. *Millers Specialist Joinery Company Ltd* v *Nobles Construction Ltd*, 3/8/2001 (E)
84. *Maxi Construction Management Ltd* v *Morton Rolls Ltd*, 7/8/2001 (S)
85. *Paul Jenson Ltd* v *Stavely Industries plc*, 27/9/2001 (E)
86. *William & Davis Oakley* v *Airclear Environmental Ltd & TS Ltd*, 4/10/2001, (E)
87. *Jerome Engineering Ltd* v *Lloyds Morris Electrical Ltd*, 23/11/2001, (E)
88. *Schimizu Europe Ltd* v *Automajor Ltd*, 17/1/2002, (TCC)(E)
89. *C&B Scene Concept Design Ltd* v *Isobars Ltd*, 14/12/2001, (Court of Appeal) (E)
90. *Karl Construction (Scotland) Ltd* v *Sweeney Civil Engineering (Scotland) Ltd*, 17/1/2002, (Court of Appeal) (S)
91. *Solland International* v *Daraydon Holdings*, 15/2/2002, (S)
92. *Total M&E Services Ltd* v *ABB Building Technologies Ltd, (formerly ABB Stewards Ltd)*, 26/2/2002, (TCC) (E)
93. *RJT Consulting Engineers Ltd* v *D M Engineering (NI) Ltd*, 8/3/2002, (Court of Appeal) (E)
94. *Edmund Nuttall Limited* v *R G Carter Limited*, 21/3/2002, EWHC 400 (TCC) (E)
95. *Balfour Beatty Construction Ltd* v *The Mayor and Burgesses of the London Borough of Lambeth*, 12/4/2002, EWHC 597 (TCC) (E)

Appendix 8

Summaries of adjudication cases

The following are summaries of some of the pertinent details of, and decisions reached in, the adjudication cases listed in Appendix 7, with additional comments in italics. For fuller information, reference should be made to the court or other reports on the case(s) concerned. It will be noted that the first three cases pre-date the Construction Act. As stated earlier, (E) and (S) in the headings indicates an English and Scottish case, respectively.

Cameron (A) Ltd **v** ***John Mowlem*****, 1990 52 BLR 24**

An Adjudicator considered matters outside his powers in order to reach a decision, i.e. he exceeded his jurisdiction. The court refused to enforce the Adjudicator's decision.

Drake & Scull Engineering Ltd **v** ***McLaughlin & Harvey plc*****, 1992 60 BLR 102**

Effective enforcement followed a disputed Adjudicator's DOM/1 sub-contract decision.

The main contractor was forced by the court following Order 14 proceedings to comply with an Adjudicator's decision to require disputed set-off money to be placed into a trustee stakeholder account.

[*A current view is that such accounts contravene the spirit of the Construction Act by delaying early payment. In fact, they may be non-compliant with the Construction Act. If so, and if expressly incorporated into contractual adjudication procedures, a Party would be entitled to use the Scheme instead! e.g. see* Allied London *v* Riverbrae, *later.*]

Mercury **v** ***Director General of Telecommunications*****, 1994 1388 SJLB 183**

An Adjudicator should decide issues other than jurisdiction, the latter being a matter that could be decided later by the courts.

Macob Civil Engineering Ltd **v** ***Morrison Construction Ltd*****, 12/2/99, BLR 93 (TCC)(E)**

This was the first case arising under the Construction Act. Morrisons engaged Macob to carry out certain ground works. The project was a 'construction contract' but the conditions of contract itself failed to comply with the minimum requirements of the Act and the Adjudicator decided to apply the Scheme. He made an award in favour of Macob (which he also made peremptory) but Morrison refused to pay and was therefore taken to court by Macob. There was some debate about the best means of enforcing the award of an Adjudicator. The court made it clear that for monetary awards it preferred an application for summary judgment and not a mandatory injunction. This is the procedure which is now generally accepted. The court advised Adjudicators not to make all their decisions automatically peremptory, as the practice at the time seemed to be, but that if they did so, the courts themselves would still decide whether or not the award should be enforced on that basis.

It was suggested that the word 'decision' in the Act should be interpreted to mean a lawful and valid decision and that where (as in this case) the decision was challenged, then it could not be regarded as lawful and valid, and could not be enforced, until the matter had been settled by arbitration. A sist (a temporary halt to legal proceedings) under Section 9 of the Arbitration Act was sought. This was not accepted by the court, which furthermore held that the decision of an Adjudicator, where its validity is challenged and even where it is mistaken on the facts, is nevertheless to be properly interpreted as a decision which is binding and enforceable.

Rentokil Ailsa Environmental Ltd v *Eastend Civil Engineering Ltd*, 12/3/99 and 31/3/99, CILL 1999 1506 (S)

Arrestment is a Scottish principle by which money lodged with a third party can be frozen pending the outcome of proceedings. In this case, following an adjudication award against it, Rentokil deposited an amount including the amount of an adjudication award with Eastend's solicitors but lodged an arrestment preventing it from being actually given to Eastend. On appeal, it was decided that a Party that had participated in adjudication and then embarked on legal proceedings covering the same subject matter could not bypass the adjudication decision using such legal proceedings. Further, it was stated that such attempts would be considered vexatious by the courts. As the arrestment was intended to avoid the Adjudicator's award it was refused.

Outwing Construction Ltd v *H Randell & Son Ltd*, 15/3/99, BLR 156 (TCC)(E)

The contract was a DOM/1 sub-contract. The Adjudicator decided that its particular adjudication terms did not comply with the Construction Act and that the Scheme was applicable. The Adjudicator decided a dispute concerning money in favour of Outwing, ordering payment within 7 days, stating that compliance was peremptory, thus permitting Outwing to apply to the courts, if necessary, to enforce payment. Payment was not forthcoming. Outwing issued an invoice for the amount requiring payment in 7 days, issued proceedings for the amount claimed plus costs (the action to be stayed on payment of the full amount into court), obtained leave to issue a summons for summary judgment and additionally asking that the usual period allowed in debt recovery proceedings by the courts for a response by the other Party be substantially shortened. The courts agreed. However, just before the actual hearing took place before a judge for a decision, a cheque had been paid for the amount of the Adjudicator's award. Outwing then wanted reimbursement of the additional costs it had incurred in seeking a court order to force payment. Despite it being argued that, as the money covering the Adjudicator's decision had been paid there was no longer an issue with the courts, the courts disagreed, ordering payment of the additional costs.

The judge confirmed that whilst payment automatically stayed the proceedings, in such circumstances the court could lift the stay in order to decide entitlement to costs.

[*In summary, if one Party tries brinkmanship and delayed payment of amounts due at the last possible moment, then when it comes to court, even though the principal sum has been paid, interest and costs may still be recovered.*

This case is another example of the courts supporting an Adjudicator's decision, showing also that the courts will support a fast track procedure (even faster than the normal timetable for the enforcement of payment of any other debt) that leads to earlier enforcement of an adjudication decision, if appropriate. This would be the case where there is a straightforward adjudication decision, but it might not always be so, for example, where a decision is disputed on the grounds of an Adjudicator exceeding its jurisdiction, not acting impartially, an irregularity in the handling of the adjudication and so on.]

A Straume (UK) Ltd v *Bradlor Developments Ltd*, 7/4/99, CILL 1999 1520 (Chancery Division)(E)

Following Bradlor going into administration, the administrator served two Notices of Adjudication on Straume to recover outstanding money under the pre-existing JCT 1980 contract. Straume countered with adjudication notices of its own covering alleged set-off. The issue went to court, where it was decided that adjudication is a quasi-legal process falling within Section 11 (3) of the Insolvency Act 1986 (which basically states that no legal proceedings can be brought against a company in administration or administrative receivership without leave of the court). As a consequence, the leave of the court was necessary before adjudication could be commenced. In this case, such leave was refused.

Leave to appeal was granted to Straume in respect of both the decision and the refusal to grant leave. However, the other adjudications involving Straume resulted in Straume gaining the off-set amounts they wanted, so no appeal was necessary.

A & D Maintenance and Construction Ltd v *Pagehurst Construction Services Ltd*, 23/6/99, CILL 1999 1518 (TCC)(E)

Pagehurst's contract was terminated by the Employer and Pagehurst did the same to their sub-Contractor A & D Maintenance (with whom there was no written contract as such, although there was a 'sub-contract order', an exchange of written submissions in arbitration, both Parties proceeded as if there was a contract and neither Party denied a contract was in place). They refused to pay them outstanding money on the grounds of poor progress and defective work. On the basis that A & D Maintenance considered there was a contract and that contract did not have any adjudication procedures and hence, as the option to use the Scheme applied, A & D Maintenance submitted their claim to adjudication.

An Adjudicator decided in favour of A & D Maintenance. The latter then applied for summary judgment to enforce the award. It was decided *inter alia* that:

1. Under the definitions in the Construction Act (i.e. a reference in legal proceedings to a contract which is not denied is treated as written evidence of a contract) there was 'an agreement in writing' between Pagehurst and A & D Maintenance, despite there being no written contract.
2. In the absence of express adjudication provisions, the Scheme applied.
3. A Notice of Dispute could be given 'at any time', even though the contract had been terminated by the time the Notice was given.
4. Pagehurst had not adequately shown other reasons why the case should be decided by further litigation and not by summary judgment, and agreed to the latter.

[*This case followed the trend of enforcement set in Macob. No proper jurisdictional objection was made and therefore the court refused to consider the merits of the Adjudicator's decision and simply enforced that decision by summary judgment.*]

Allied London & Scottish Properties plc v *Riverbrae Construction Ltd*, 12/7/99, CILL 1999 1541, BLR 346 (S)

An Adjudicator ordered payment of an award within 14 days. It was claimed that there was uncertainty about the financial standing of the 'winning' Party (which was not substantiated). It was submitted by the 'losing' Party that although a number of alternatives such as placing the money awarded on deposit in the joint names of the Parties had never been put to the Adjudicator, the Adjudicator ought of his own volition to have considered them. The court disagreed that an Adjudicator was under any such obligation.

The Project Consultancy Group v *The Trustees of the Gray Trust*, 16/7/99, CILL 1999 1531, BLR 377 (TCC)(E)

An Adjudicator decided that a contract existed that came into existence after 1 May 1998, i.e. the Construction Act applied and the Adjudicator had jurisdiction to make an award. The Adjudicator awarded payment of outstanding fees to the Project Consultancy Group. The latter applied for summary judgment under CPR24.

Any contract formed prior to 1 May 1998 does not come under the Construction Act. There are thus no applicable statutory adjudication provisions and hence there can be no binding adjudication decision based on those provisions.

The Trustees had always made it clear that they disputed the Adjudicator's jurisdiction on the foregoing grounds even though they continued to participate in the adjudication on a without prejudice basis.

[*Whether the former should make any difference is not clear, but it helped the court decide in this case. It appears to be a useful stance to take.*]

In fact, the court stated that the issue of when, if at all, in fact and/or law, a contract had been concluded was unclear without full evidence and argument and this had not occurred. The matter was thus not a straightforward one suitable for summary judgment.

[*There are several clear points arising from the court's judgment:*

1. *An Adjudicator must be sure that a valid contract is in place before commencing a statutory adjudication under the Construction Act.*
2. *A Party challenging, or contemplating challenging, jurisdiction should make that clear to the Adjudicator and other Party, but nevertheless continue to participate in the adjudication [not least so that if a jurisdiction decision goes against the Party, they are not left with the consequences of (a probably more adverse) ex parte Adjudicator's decision].*
3. *Although the Parties, say, by it being incorporated within adjudication procedures, can agree that an Adjudicator can decide on certain matters of jurisdiction, where that jurisdiction is challenged on the grounds that it is outside the Parties' agreement, it is for the courts to decide (not the Adjudicator) after application by the challenging Party. Similarly, where there is no agreement to let an Adjudicator decide on a specific jurisdiction issue, then any adjudication decision on that matter could be challenged. In other words, an Adjudicator cannot decide on its own jurisdiction unless permitted by the Parties.*
4. *The courts will always consider any defence submitted to any claim decided by an Adjudicator, before agreeing to summary judgment.*

This was the first case to decide that not all Adjudicator's decisions would be automatically enforced. In summary, drawing a distinction between his decision in Macob, the same judge held that if a proper jurisdiction challenge could be made out (in this case that the contract pre-dated the Act coming into force), then a challenge could be mounted against an application for enforcement by summary judgment. In fact, the date of the contract was surrounded by such confusing facts that the judge refused to determine the matter summarily and put the case out for a full hearing.]

John Cothliff Ltd v *Allen Build (North West) Ltd*, 29/7/99, CILL 1999 1530, BLR 426 (Liverpool County Court)

Although not expressly permitted, the Construction Act and original Scheme do not prevent an Adjudicator from awarding costs (although it will be noted that the proposed amended Scheme for England 2001 expressly excludes an Adjudicator from considering costs).

In this County Court case (which does not therefore create formal legal precedent), the court confirmed that the Adjudicator had such power when payment of costs were expressly requested by one of the Parties, and arguments for and against had been presented to the Adjudicator. That part of the adjudication decision was as binding as the rest of the decision. However, the court stated that if costs were not mentioned by the Parties then an Adjudicator should not raise the issue itself.

[*It will be noted that some adjudication procedures, e.g. CIC, ICE and TeCSA expressly exclude an Adjudicator deciding on costs, so the foregoing would not apply.*]

Palmers Ltd v *ABB Power Construction Ltd*, 6/8/99, CILL 1999 1543, BLR 426 (TCC)(E)

Palmers were scaffolding sub-Contractors for the erection of scaffolding for a heat recovery steam generator boiler at a power generation site. A dispute arose which Palmers referred to adjudication. A key initial argument between the Parties was whether a contract for the erection of scaffolding in such circumstances was a 'construction contract' under the Construction Act. The Parties went to the court to obtain an quick decision on this jurisdiction matter.

The court decided that an Adjudicator could not decide its own jurisdiction, as that was a matter for the court. It also decided that although much work on an power generation project was excluded from the Construction Act, others were not. In fact, scaffolding which was preparatory to an excluded operation, may itself be included. Hansard supported a conclusion that scaffolding on a power generation site was included.

[*This fast track procedure should be favoured by all Parties and an Adjudicator as it prevents wasting time and money and eliminates future uncertainty.*]

Lathom Construction Ltd v *Anne & Brian Cross*, 29/10/99, CILL 1999 1568 (TCC)(E)

A dispute was referred to adjudication but the Parties settled before the decision was made. A dispute later arose due to conflicting understandings as to what was agreed in the settlement and one Party referred the dispute back to the Adjudicator on the basis that the claim arose originally from a construction contract. The other Party refuted that, stating that the settlement agreement was not itself a construction contract and so any related subsequent dispute should not have been referred to adjudication on that basis. [*It will be noted that would not be the case if the settlement agreement itself contained adjudication provisions.*] Apparently, the Adjudicator found that a settlement existed between the Parties but nevertheless went on to decide the matter. The other Party therefore disputed an application for enforcement by the court of the Adjudicator's decision. The court held that a proper settlement disentitles the Parties from an adjudication.

[*From this it will be apparent that the existence of a settlement can be used in an attempt to stop an adjudication or be a defence to adjudication decision enforcement proceedings.*]

Homer Burgess Ltd v *Chirex (Annan) Ltd (1)*, 10/11/99, CILL 2000 1580 (S)

It was decided that the word 'plant' in Section 105 (2)(c) of the Act includes pipework linking various pieces of equipment and hence the latter was not a construction operation. The Adjudicator did not, therefore, have jurisdiction to deal with a dispute concerning that pipework.

An Adjudicator's decision concerning such a jurisdictional issue is a preliminary non-binding decision which can be reviewed by the courts, if it is claimed that the decision was based on the Adjudicator making an error of law. [*The position where the decision was based on the Adjudicator making an error of fact is not dealt with by this case.*] In this Scottish case, it was also decided that in Scotland Adjudicators are expected to investigate their own jurisdiction where it is objected to.

[*This case has turned out to be a significant case in adjudication although it may be drawing a distinction between the positions in Scotland and in England. On the facts, the case is another interpretation of the application of the 'construction operations' and 'construction contract' provisions in the Act. However, the case has wider importance. As often happens when seeking to understand a court decision, a number of commentators had seized in isolation upon a number of remarks made in Macob and used them to suggest that the court had held that even a breach of the rules of natural justice would not prevent the courts from enforcing Adjudicator's decisions (although subsequent judges in England have refuted this). In this case the court decided that, in Scotland at any event, an error of law (including a breach of the Rules of Natural Justice) would be a valid jurisdictional objection to enforcement. This approach has been subsequently endorsed in a number of English decisions.*]

Bouygues (UK) Ltd v *Dahl-Jensen (UK) Ltd (1)*, 17/11/99, BLR 522, CILL 1999 1566 (TCC)(E)

A CIC Model Procedure adjudication arose out of a sub-contract involving large claims and counter-claims. Dahl-Jensen went into voluntary liquidation before the Adjudicator's decision was reached. It was a reasoned decision supported by calculations which awarded money to Dahl-Jensen and ordered Bouygues to pay the full Adjudicator's costs. Bouygues refused to pay, stating that there was a mistake in the calculations calculating the money due, and asked the TCC either to rule that the decision was void and should be set aside or, alternatively, to remit the decision to the Adjudicator for review. In defence, Dahl-Jensen argued that if there was an error it was not on an issue that was outside the Adjudicator's jurisdiction (i.e. it was on matters that had been referred to the Adjudicator for a decision), and requested summary judgment for the amount awarded and reimbursement of their already paid portion of the Adjudicator's costs.

The judge agreed with Dahl-Jensen. It appears that unless an issue that lay outside an Adjudicator's jurisdiction was involved, a decision merely wrong on a point of law, fact or in calculations would nevertheless be supported, even though occasionally (say, due to insolvency of the other Party) a Party was not able to try to recoup losses in later litigation or arbitration.

The Adjudicator had made a mathematical error that had almost reversed what should have been the correct decision, but the Adjudicator had answered the right question in the wrong way. The issue was whether that should be enforced or corrected. The court decision at first instance, by analogy with expert determinations, established for adjudication the principle that if the Adjudicator has answered the right question in the wrong way, his decision will be binding but if he has answered the wrong question then his decision will be a nullity.

[*This was the first adjudication case to go to the Court of Appeal, which was not entirely convinced of the basic principle but declined to interfere with the first instance decision. However, on the basis of a point relating to insolvency law which had not been raised in the first instance court or raised before them, the Court of Appeal granted a stay of execution – see later.*]

Homer Burgess Ltd v *Chirex (Annan) Ltd (2)*, 18/11/99 (S)

This is the second part to the earlier case. The court decided that where it identified an error in law in part of the Adjudicator's decision, it was up to the court to decide whether the whole decision was therefore void or if there were some parts that could be supported. The court also considered whether it or the Adjudicator should correct any error in the decision. In this case [*which by this time, of course, was well past the 28-day time limit for the Adjudicator's decision*], the decision was sent back to the Adjudicator to amend.

[*It will be noted that this might well now be a valid approach in Scotland but it is by no means certain that the same approach is a valid one in England.*]

Sherwood & Casson Ltd v *MacKenzie Engineering Ltd*, 30/11/99, CILL 2000 1577 (TCC)(E)

This was related to a Scheme adjudication which requires an Adjudicator to resign where the dispute referred to the Adjudicator is 'the same or substantially the same as one which has previously been referred to adjudication'.

It was confirmed by the court that in this case, a dispute on the final account (which involved some revaluation of work done) was different from a dispute on an earlier interim account, despite common elements.

[*It appears that the facts of each particular case will be important, including the amount of additional information made available by the Parties for final evaluation, and perhaps the extent of any revaluation.*]

Under the Scheme an Adjudicator has the express jurisdiction to determine whether or not two such disputes are substantially the same. This does not, however, prevent the courts from assessing the decision by the Adjudicator, but that will require substantial grounds to be presented supporting the view that the Adjudicator was wrong.

The judge summarised previous decisions on the enforcement of adjudication decisions:

1. A decision challenged because of factual, legal or procedural error will be enforced (*Macob*).
2. A wrong decision will be enforced (*Bouygues*).
3. A decision may be challenged due to lack of empowerment of an Adjudicator by the Act because there is no construction contract (*Project Consultancy*) or the Adjudicator has gone outside its terms of reference (*Bouygues*).
4. Mistakes will inevitably occur in a speedy process, so care is needed not to confuse wrong decisions as necessarily being an excess of jurisdiction (*Bouygues*).

5. Whether a contract has come into existence (hence affecting the jurisdiction of an Adjudicator) will be determined by the courts on a balance of probabilities, after considering any necessary oral or written submissions (*Project Consultancy*).

[*The classification of jurisdictional error set down in this case has subsequently been adopted in a number of English and Scottish cases.*]

Fastrack Contractors Ltd v *Morrison Construction Ltd and Imregio UK Ltd*, 4/1/2000, CILL 2000 1589, BLR 168 (TCC)(E)

It was alleged that there was no current dispute to be adjudicated, and so the Adjudicator had no jurisdiction.

The court stated that a dispute arose once it 'had been brought to the attention of the opposing Party and that Party has had the opportunity of considering and admitting, modifying or rejecting' it. A rejection could be considered to have arisen where a Party refused to answer a claim. A dispute could also arise where there was 'a bare rejection of a claim to which there was no discernible answer in fact or in law'.

[*Thus, for a dispute to arise there must first be a claim and then a rejection, where silence for an unreasonable period can also be a rejection.*]

A dispute could be in the form 'what sum is due?' without the need for any particularised or finalised amount being stated. In this case, the Referring Party had partly particularised its claim, but widened it to include 'or any such other sums as the Adjudicator shall find payable'.

For *statutory* adjudications the Act allows only a single dispute to be referred to adjudication at any one time, but a number of issues can constitute a single dispute, i.e. a 'single dispute', as stated by the Act, constitutes all issues at any one point in time which the claimant decides to refer to adjudication.

VHE Construction plc v *RBSTB Trust Co Ltd*, 13/1/2000, CILL 2000 1592, BLR 187 (TCC)(E)

Deductions, in this case liquidated damages, can only be made from an application for payment if a written withholding notice (in accordance with Section 111 of the Act) with adequate details is provided within the required timescale, in this case 5 days before the final date for payment.

The court considered that Section 111 (1) meant that set-offs could only be considered in a claim [*e.g. an adjudication*] if an effective notice of intention to withhold payment was made. The court further agreed that additionally, to be included in an adjudication, the withholding notice had to precede the reference to adjudication. [*The court, in Northern Developments (Cumbria) Ltd* v *J & J Nichol (see following), expressly agreed with this position.*]

As the notice was not provided as required, the court ordered payment of the full amount claimed with no allowance for set-off of the liquidated damages.

[*It will be noted that the liquidated damages could be deducted later if the proper notice was produced before the next payment was due (assuming more money was going to be claimed).*]

Northern Developments (Cumbria) Ltd v *J & J Nichol*, 24/1/2000, CILL 2000 1601 (TCC)(E)

This was a Scheme dispute in which there was no express power given to the Adjudicator to award costs. However, the Parties can agree to alter the Scheme provided they do not detract from the minimum requirements of the Scheme or the Act.

Where both Parties ask for costs and do not suggest that an Adjudicator has no power to award them, the courts will assume that the Adjudicator has been given jurisdiction by the Parties. Generally, however, an Adjudicator does not have jurisdiction (unless expressly stated).

The court suggested that an alternative approach for a Party that does not want costs awarded, but is unsure what an Adjudicator might do, is to say to the Adjudicator 'I have only asked for costs in case you decide you have the jurisdiction to award them, but I submit you have no jurisdiction to make such an award.'

There is a detailed discussion on the effects of contract repudiation.

[*It will be noted that this case applies to original statutory Scheme, not contractual, adjudications. It may be overtaken by later amendments to the Scheme.*]

Samuel Thomas Construction Ltd v *Bick & Bick Ltd (t/a J & B Developments)*, 28/1/2000 (High Court Exeter)

The Construction Act is not intended to apply to a contract involving work principally on a dwelling that one of the Parties occupies or intends to occupy as a residence.

An adjudication arose and the jurisdiction of the Adjudicator was disputed on the basis of the foregoing. However, the court decided that it was not possible to say that the Works involved in total were principally a residence to be occupied by one of the Parties and so adjudication was permitted.

F W Cook Ltd v *Shimizu (UK) Ltd*, 4/2/2000 (TCC)(E)

This was a case of an Adjudicator dealing with several sub-disputes as part of a dispute referred to adjudication and there being a resulting disagreement over the meaning of the decision. Apparently, rather than obtaining clarification from the Adjudicator, the court was asked to provide its interpretation.

One of the Parties suggested that the Adjudicator's decision meant that they were entitled to a certain sum of money. The other Party argued that the Adjudicator had merely arrived at a decision on a matter of principle which, when applied, resulted in a different sum. The court held that an Adjudicator was entitled to state the decision as one of principle only and that that was, in fact, what the Adjudicator had been asked to do and had done in this case.

Workplace Technologies plc v *E Squared Ltd and Anr*, 16/2/2000, CILL 2000 1607 (TCC)(E)

A Party went to adjudication to obtain a decision on an amount it thought it was due. The other Party considered that the contract came into effect before the Act became effective, i.e. 1 May 1998, and went to the court for a declaration that the Construction Act did not apply, or a declaration on what was the true date of the contract and an injunction to stop the adjudication. The application proceeded very quickly and after hearing submissions, the court decided the contract was post 1 May 1998 and hence the adjudication was valid.

[*This is a good example of the rapid clarification of a jurisdiction issue before unnecessary time and money was spent.*]

The court indicated that, as a matter of law, injunctions will almost never be granted to stop a Party proceeding with an invalid adjudication. This must be prevented by other means such as a declaration.

Atlas Ceiling & Partition Company Ltd v *Crowngate Estates (Cheltenham) Ltd*, 18/2/2000 (TCC)(E)

A letter of intent was issued in December 1997, but it was made clear in correspondence that it was the express intention of the Parties that the contract (or sub-contract in this case) was not to be *entered into* until various outstanding issues were resolved. Work in fact commenced in January 1998 on the basis of the letter of intent, but correspondence supported a view that a contract had still not been *entered into* and that the outstanding issues were not resolved until April 1999. A dispute subsequently arose which was submitted to adjudication, but jurisdiction was disputed on the basis that the contract came *into effect* before the Act became effective, i.e. 1 May 1998. The Adjudicator decided that the sub-contract was *entered into* after 1 May and that the Act complied (i.e. the Adjudicator decided its own jurisdiction) and reached a money decision. The 'losing' Party asked the court for a declaration that there was a lack of jurisdiction. The 'winning' Party asked for summary judgment of the decision on the basis of the second Party having no real chance of succeeding with its request.

The court ordered the summary judgment to be turned into a trial of the issues, all within 18 days of court proceedings commencing. It was found that the evidence, including the express intentions of the Parties at the time of the letter of intent, showed that sub-contract was only actually *entered into* after the Act came into force and so the Adjudicator had jurisdiction. The court said that this remained the case even if the sub-contract *had effect* from the date of the letter of intent, it was not *entered into* for the purposes of the Act until April 1999. The losing Party was ordered to pay the winning Party's costs of the proceedings as assessed at the trial.

[*The better approach is for an early decision on jurisdiction by the courts, or if there is arbitration in the contract, by arbitration (and the RICS now have a rapid response procedure in place for the latter). This is a good example of the court moving quickly. Furthermore, since a declaration was granted as opposed to an injunction, the issue was (subject to any appeal) legally determined once and for all.*]

Nolan Davis Ltd v *Steven Catton*, 22/2/2000, unreported (TCC)(E)

The Architects & Surveyors Small Works Form includes adjudication. A dispute arose but it was found that a Party named in the contract did not exist. Advice received during the dispute was that based on the Companies Act 1985, i.e. if a contract is purportedly made by a company that turns out not to exist at that time, then the contract is with the person purporting to act for that company, and that person is personally liable instead.

Both Parties, one of whom was the individual just described, instructed an Adjudicator to decide the dispute, including whether there was a company to stand in place of the individual. It couldn't and found the individual was liable for a certain amount of money. The individual then asked the court to agree that there was no contract and thus no jurisdiction to adjudicate. The court refused, pointing out the joint agreement between the Parties to accept the Adjudicator's decision, which included in effect the Adjudicator deciding its own jurisdiction.

The court also said the Adjudicator could award costs because both Parties had asked for them, and refused the paying Party a 'breathing space' before paying the money.

Grovedeck Ltd v *Capital Demolition Ltd*, 24/2/2000, CILL 2000 1604, BLR 181 (TCC)(E)

There was a dispute concerning the existence or otherwise of a construction contract under the Act at the time of referral to an Adjudicator. It was agreed there were oral agreements between the Parties, but it was always refuted by the Responding Party that these were intended to create a written contract under the Act. However, an Adjudicator was appointed to decide this issue, i.e. effectively to decide on its own jurisdiction. It was only after its jurisdiction was confirmed by the Adjudicator that exchanges of written communications were sufficient to confirm that there was then a construction contract in place. The court agreed that there was no construction contract at the time of Adjudicator appointment and so no jurisdiction existed.

Another aspect was that the Scheme precluded an Adjudicator dealing with related disputes on more than one contract at the same time without agreement of the Parties. In this case there were two contracts included in one referral notice, but clearly no agreement by one of the Parties to deal with them at all, let alone at the same time.

[*This case should be looked at in the context of a number of subsequent cases on what amounts to construction contracts being in 'writing'.*]

Absolute Rentals Ltd v *Gencor Enterprises Ltd*, 28/2/2000, CILL 2000 1687 (E)

A contract contained both an adjudication clause and an arbitration clause. It was argued by analogy with *Halki* that in these circumstances the court could not interfere by giving summary judgment and that the Parties had to resolve their dispute by arbitration. The court decided *Halki* did not apply to adjudications and granted summary judgment.

[*The decision of Halki Shipping Corporation v Sopex Oils Ltd (1998) 1 WLR 726 was a significant one in construction law because it established that if there was an arbitration clause then the Parties were entitled to that form of dispute resolution as an alternative to the courts (which could not then interfere).*]

Bloor Construction (UK) Ltd v *Bowmer & Kirkland (London) Ltd*, 5/4/2000, CILL 2000 1626, BLR 764 (TCC)(E)

A decision made within an adjudication timetable but not provided to the Parties until after the timetable had expired would be out of time. The court stated: 'In the absence of a specific agreement by the Parties to the contrary, there is to be implied into the agreement for adjudication the power of the Adjudicator to correct an error arising from an accidental error or omission or to clarify or remove any ambiguity in the decision which he has reached, provided this is done within a reasonable time and without prejudicing the other party'.

[*In other words, providing the contract does not state otherwise, an Adjudicator may correct a mathematical error or an ambiguity in reasonable time as long as it does not prejudice a Party. It will be noted that the amendments to the English statutory Scheme, 2001, expressly provides for the corrections of some errors.*]

Bridgeway Construction Ltd v *Tolent Construction*, 11/4/2000, unreported (TCC Liverpool)

The court supported an amendment to CIC adjudication clauses which stated that a Referring Party, win or lose, had to pay the Responding Party's costs.

[*This would probably not have been the case if it could have been shown that the amendment was meant to be a deterrent to adjudication, unless the amendment had been explicitly drawn to the Referring Party's attention at the time of signing the contract.*]

Tim Butler Contractors Ltd v *Merewood Homes Ltd*, 12/4/2000, unreported (TCC Salford)(E)

An Adjudicator was asked by a Party to decide whether the terms of a contract included an entitlement to staged/interim payments, and if so, to adjudicate and decide a dispute on outstanding money. The other Party said that it was agreed between the Parties that the duration of the Works was less than 45 days and so there was no entitlement, nor was there a construction contract to which the Scheme should apply. The Adjudicator decided that there was no agreement on the duration of the Works and so adjudication could proceed. It decided in favour of the first Party.

The court stated the Adjudicator was entitled to make a decision on whether there was a construction contract or not (whether it was decided rightly or wrongly). This was a decision on a dispute relating to the terms of a contract. Even if it made a wrong decision, the Adjudicator was entitled to reach such a decision as the conditions of contract provided for staged payments and Paragraphs 6 and 7 of the Scheme described how payment should be made.

Edmund Nuttall Ltd v *Sevenoaks District Council*, 14/4/2000, unreported (E)

The Adjudicator decided that it had no power to amend an error in its decision. The court decided differently and that the decision could be corrected within a reasonable period.

Herschel Engineering Ltd v *Breen Property Ltd (1)*, 14/4/2000 (TCC)(E)

A Party decided to sue in the local County Court for money owed to it instead of using, say, adjudication. As that process was slow, the Party then decided to try adjudication. The other Party refused to participate and the Adjudicator awarded against it. The other Party refused to pay and the first Party sought enforcement through the Court. In defence, the other Party asked the Court to decide on the matter that it was now involved in two venues at the same time for the same dispute – simultaneous litigation and arbitration would not be permitted, so why should litigation and adjudication?

The Court said that simultaneous litigation (or arbitration) and adjudication was permitted as it was not expressly excluded by the Act, adjudication was not binding, and the Act expressly permitted adjudication at any time.

The other Party asked for payment to be delayed as litigation was proceeding. In fact it had not commenced and no date had been set. The Court refused, but said it might have reached a different decision if there was real doubt about the ability of the first Party to refund money if subsequent litigation went against it.

[*Running the same dispute side-by-side in parallel proceedings is normally frowned upon, but not in the case of adjudication. It is therefore possible to have an adjudication and a parallel arbitration (or litigation) running at the same time covering the same dispute. It is even, theoretically, possible to have both arbitration and litigation running at the same time as adjudication, although one of the former is normally likely to be sisted (i.e. temporarily halted).*]

Strathmore Building Services Ltd v *Colin Scott Greig (t/a Hestia Fireside Design)*, 18/5/2000 (S)

Notice under Sections 110 and 111 of the Construction Act must be in writing, be specific and not pre-date the demand for payment.

[*This was not, strictly speaking, an adjudication case – there was no adjudication but the court's comments are considered relevant.*]

John Mowlem & Company plc v *Hydra-Tight Ltd (t/a Hevilifts)*, 6/6/2000, CILL 2000 1650 (TCC) (E)

Both Parties agreed that adjudication procedures that incorporated 'matters of dissatisfaction' provisions were unlawful because they were not compliant with the 'at any time' phrase used in the Construction Act.

[*It will be noted that 'matters of dissatisfaction' provisions are not unlawful, they are only non-compliant with one of the Construction Act's requirements. Contractual adjudication procedures that are not compliant with the Act can still be used if a Referring Party does not want to use its entitlement.*]

R G Carter Ltd v *Edmund Nuttall Ltd*, 21/6/2000 (E)

A contract contained a clause obliging the Parties to have a mandatory mediation prior to any other dispute resolution procedures, including adjudication. The judge held that this was non-compliant with the Construction Act entitlement to adjudication 'at any time'.

Stiell Ltd v *Riema Control Systems Ltd*, 23/6/2000 (S)

This was a Scottish case involving valid arrestments. It was the first to go to the Court of Appeal in Scotland. The question was whether, with the Adjudicator having arrived at his decision, those arrestments should be restricted to a lesser sum. The answer, apparently, was no.

Christiani & Neilsen Ltd v *The Lowry Centre Development Company Ltd*, 29/6/2000, unreported (TCC)(E)

This case involved, *inter alia*, the rectification of a deed. It was decided that the Adjudicator did not have ad hoc power to determine his own jurisdiction but in this case the rectification of deeds was something that could be within the jurisdiction of an Adjudicator. The court held that the Adjudicator's decision was correct.

Nottingham Community Housing Association Ltd v *Powerminster Ltd*, 30/6/2000 (TCC)(E)

The work of repairing, dismantling and maintaining domestic gas appliance in a building is a construction activity coming under the Construction Act.

KNS Industrial Services (Birmingham) Ltd v *Sindall Ltd*, 17/7/2000 (TCC)(E)

What matters in an adjudication is the Notice of Adjudication and that the dispute cannot later be adjusted by the Referral Notice.

An Adjudicator's powers are limited to those conferred by the contract and thus no more than those of a contract administrator such as an architect, engineer or surveyor when entrusted with the resolution of disputes.

If an Adjudicator allows a deduction where there should have been, but wasn't, a valid withholding notice, that would not be an excess of jurisdiction but an error. The decision would still be supported by the court.

Ken Griffin and John Tomlinson (t/a K & D Contractors) v *Midas Homes Ltd*, 21/7/2000, unreported (E)

An adjudication took place in relation to a number of invoices and a decision was made which it was then sought to enforce in the courts. No proper time had been given to the Responding Party to consider some of these invoices with the Referring Party prior to the adjudication and some of the invoices had not been properly included in the Notice of Adjudication. The court only enforced the award in relation to the invoices that had been discussed and validly included.

Shepherd Construction Ltd v *Mecright Ltd*, 27/7/2000, BLR 489 (E)

The court held that Section 105 of the Construction Act required to be narrowly construed, e.g. that statutory adjudication is available only in respect of those disputes arising *under* the contract and not *in connection with* or *arising out of* the contract. The latter would include settlements. Accordingly, it was held an Adjudicator had no power to consider the validity of a settlement which stood until such time as it was set aside by the court or the Parties.

Mitsui Babcock Energy Services Ltd, Petitioner (Judicial Review), 27/7/2000 (S)

Equipment and pipework prefabricated elsewhere for installation in mechanically otherwise complete heat recovery steam generating boilers are not covered by the Construction Act where the primary activity on site was petrochemical works, an activity excluded by Section 105 (2)(c).

Herschel Engineering Ltd v *Breen Properties Ltd (2)*, 28/7/2000, unreported (TCC)(E)

A stay in judgment was requested on the grounds that the other Party could not repay money if a then existing litigation reversed an original decision that they should be given that money. The court accepted that although the basis of the argument presented was valid, in this case insufficient grounds had been provided to show that the other Party would not be able to repay the money.

Bouygues (UK) Ltd v *Dahl-Jensen (UK) Ltd (2)*, 31/7/2000, CILL 2000 1616, 1 BLR 49 (Court of Appeal) (E)

The earlier first instance decision of the court concerning allowing summary judgment even though an Adjudicator had answered the right question the wrong way was supported. Under the Insolvency Rules 1986, rule 4.90, a Party was entitled to set off any cross-claim against the other, insolvent, Party. Unfortunately, set-off had not been raised in the initial case and so again there had been no reason to refuse summary judgment.

ABB Power Construction Ltd v *Norwest Holst Engineering Ltd*, 1/8/2000 (TCC) (E)

Installation works at a power station were not 'construction operations' under the Construction Act and there was no entitlement to an adjudication.

George Parke v *The Fenton Gretton Partnership*, 2/8/2000 (E)

The court decided that an Adjudicator's decision, behind which a court will not go, may form the basis for a statutory demand. However, the court also held that, depending on the facts, a valid counter-claim, set-off or cross-claim might be sufficient to allow the statutory demand to be set aside, as it was in this case.

Whiteways Contractors (Sussex) Ltd v *Impresa Castelli Construction UK Ltd*, 9/8/2000 (TCC) (E)

The court held that Section 111 of the Construction Act makes no distinction between set-off and abatement and neither can thus be taken into account by an Adjudicator unless included in a valid Section 111 Notice.

Discain Project Services Ltd v *Opecprime Development Ltd (1)*, 9/8/2000, 8 BLR 402 (TCC)(E)

The court held that, even if not referred to in adjudication rules, under common law adjudications must be conducted in accordance with the Rules of Natural Justice. Furthermore, a failure on the part of an Adjudicator to follow the Rules of Natural Justice is a jurisdictional reason for a court to decline to enforce an Adjudicator's decision. In this case, where an Adjudicator communicated with one Party only and failed to relate that to the other Party thereby creating the appearance of bias, the court refused to enforce the Adjudicator's decision.

Universal Music Operations Ltd v *Flairnote Ltd and Anr*, 24/8/2000, unreported (E)

A Consultant was involved in arranging a contract between two Parties. Faced with an adjudication, one of the Parties sought to argue that the contract in question was in fact between the other Party and the Consultant. The court held that based on the evidence the contract was between the Parties themselves.

Elenay Contracts Ltd v *The Vestry*, 30/8/2000, CILL 2000 1679 (TCC)(E)

It was held that because an adjudication was not a 'final determination', Article 6 of the European Convention on Human Rights did not apply to adjudications.

Cygnet Healthcare plc v *Higgins City Ltd*, 6/9/2000 (E)

A preliminary issue related to an adjudication had, with the agreement of the Parties, been put out to an Arbitrator for his decision. The court was persuaded to adjourn the question of the enforcement of the adjudication decision pending receipt of the decision on the preliminary issue, because otherwise the court would have to determine the very question which had been submitted to the Arbitrator.

Woods Hardwick Ltd v *Chiltern Air Conditioning*, 2/10/2000 (TCC) (E)

In this case an Adjudicator did not refer all of the information provided to him by one Party to the other Party and, unusually, produced a witness statement for a Party, the latter being considered by the court to indicate a lack of impartiality. The court decided these were a breach of the Rules of Natural Justice and declined to enforce the decision of the Adjudicator.

Maymac Environmental Services Ltd v *Faraday Building Services Ltd*, 16/10/2000, CILL 2000 1686 (TCC) (E)

No one objected to jurisdiction during the adjudication but it was brought up as a defence to enforcement proceedings. The court held that the Party was estopped by 'representation' and 'convention' from denying jurisdiction. The court supported the Adjudicator's decision.

ABB Zantingh Ltd v *Zendal Building Services Ltd*, 12/12/2000 (E)

The Parties agreed to adjourn an adjudication until a question of law could be determined by the courts relating to the primary activity on site as stated under Section 105 of the Construction Act. In this case it was decided that the primary activity was printing not power generation so that the operations in question did not fall within the exceptions of subsection 105 (2)(c). It was agreed that statutory adjudication could validly proceed.

Karl Construction (Scotland) Ltd v *Sweeney Civil Engineering (Scotland) Ltd*, 21/12/2000 (S)

An Adjudicator decided that the payment provisions of a contract were non-Construction Act-compliant and unilaterally (e.g. without giving Parties an opportunity to discuss the issues – indeed the Parties had not asked the Adjudicator to consider such matters) decided the Scheme (for Scotland) was to be used instead and proceeded to a decision. The court was 'left feeling uncomfortable' but decided that although there was a procedural 'mistake' there was no excess of jurisdiction. (See also subsequent Court of Appeal decision, 17/1/2002.)

LPL Electrical Services Ltd v *Kershaw Mechanical Services Ltd*, 2/2/2001 (E)

An Adjudicator interpreted a contract to mean that he could consider wider issues than a specific amount referred to him to decide. It was decided that an error of law or interpretation was not outside jurisdiction. It was also decided that no sanction applied to a failure to issue a Section 110 Notice.

Holt Insulation Ltd v *Colt International Ltd*, 2/2/2001 (Liverpool TCC)

A Scottish company and an English company entered into a contract to do Works in England. The Parties fell into dispute. An initial adjudication was completed and a decision made. Then there was a second adjudication (but this time under objection) and a decision was made. The Scottish company anticipated opposing enforcement proceedings in England on the basis of a jurisdictional objection. However, the English company, in whose favour the award was, chose simply to issue a Statutory Demand against the Scottish company in Scotland. The matter came before a Scottish court which granted an adjournment to allow the Scottish company to see if they had a defence. The Scottish company commenced proceedings against the English company in the English TCC, seeking a declaration based upon their objection to jurisdiction, which was that the dispute referred to the Adjudicator in the second adjudication was the same as the dispute referred to the Adjudicator in the first adjudication and this was not allowed by the Scheme. The TCC held that in this case, although arising out of the same matter, the disputes were not the same.

Glencot Development and Design Co Ltd v *Ben Barrett & Son (Contractors) Ltd*, 2/2/2001 and 13/2/2001 (E)

The Parties met at a hearing at the commencement of an adjudication and nearly reached an agreement. They asked the Adjudicator to act as a Mediator to help them reach a settlement. He did so but no settlement was reached. The Adjudicator, aware of the potential problems, after discussions with, and initial agreement by, the Parties then carried on with the adjudication. One of the Parties later withdrew its agreement and used the fact of the Adjudicator's confidential discussions with the Parties in the mediation as a reason for claiming possible unconscious bias in the Adjudicator's decision. The court, whilst not being critical of the Adjudicator, agreed.

Rainford House Ltd (In Admin Receivership) v *Cadogan Ltd*, 13/2/2001 (E)

This case involved the administrative receivership of the 'winning' Party to an Adjudicator's decision. Of concern is whether recovery of money paid was a possibility if the dispute should go to litigation or arbitration and the Adjudicator's decision be reversed. The court stated that they would not normally grant summary enforcement of an Adjudicator's decision in such circumstances. In this case the court supported the application for summary judgment against the 'losing' Party but required that the money in question was paid into court.

Staveley Industries plc (t/a EI.WHS) v *Odebrecht Oil & Gas Services Ltd*, 28/2/2001 (E)

It was decided that structures being constructed for use in the oil industry below the low-water mark did not constitute works or structures 'forming part of the land' as required by Section 105 (1) of the Construction Act.

Joseph Finney plc v *Gordon Vickers and Gary Vickers (t/a The Mill Hotel)*, 7/3/2001 (E)

The Parties reached a settlement which included an agreement not to adjudicate. There was a dispute as to whether the agreement was binding and, on the basis that it was not, one Party with a dispute commenced adjudication. The question arose as to whether it was correct to use arbitration to resolve whether the agreement was binding or not. The court's finding was that the option of arbitration, although it could have covered the question of the settlement, had not been successfully incorporated into the JCT contract. This left litigation as the remaining option. The court held that the settlement agreement had been repudiated (presumably by one Party not completing its side of the agreement) so the Parties were free to adjudicate.

Watson Building Services Ltd, Petitioner (Judicial Review), 13/3/2001 (S)

The Parties requested an Adjudicator to determine a dispute over the interpretation of the terms of a sub-contract between them, which included issues relating to its appointment and hence jurisdiction.

On the facts of this case it was agreed that the Parties had delegated to the Adjudicator the right to determine its jurisdiction and were bound by that decision and the court was not prepared to interfere.

The court confirmed the approach adopted in *Homer* that in Scotland an Adjudicator is expected to investigate its own jurisdiction upon an objection being made.

The Scottish court at first instance also adopted for Scotland the categorisation of jurisdictional objections set out in *Fastrack*. It also stated that reference to a main contact within a sub-contract does not import into that subcontract the dispute resolution procedures of the former.

Austin Hill Building v *Buckland Securities Ltd*, 11/4/2001 (E)

Although the whole dispute resolution process including final resolution by the courts might comply with the European Convention on Human Rights, through the Human Rights Act, the latter did not apply to adjudications. An Adjudicator is not a public authority nor a 'tribunal' before whom legal proceedings can be brought, nor need an Adjudicator hold a public hearing. In any case, where a Party does not ask for the latter, he waives his right, if any, to it.

Discain Project Services Ltd v *Opecprime Development Ltd (2)*, 11/4/2001 (E)

An Adjudicator admitted that it had telephone discussions with one of the Parties where details relevant to the dispute were discussed. However, details of the discussions were not given, or not given until later, to the other Party nor were their views sought. Thus a fair-minded and informed observer might conclude that there was a real possibility of bias by the Adjudicator.

A court will not enforce an Adjudicator's decision where there had been a substantial breach of the Rules of Natural Justice by the Adjudicator. The Adjudicator must undertake the adjudication as fairly as any limitations imposed by Parliament permit.

Farebrother Building Services Ltd v *Frogmore Investments Ltd*, 20/4/2001 (E)

This was a contractual adjudication under the TeCSA adjudication rules. Those rules expressly give an Adjudicator the power to rule upon its own jurisdiction and this Adjudicator did so. The court held that his decision in that respect was binding on the Parties.

The court also held that an Adjudicator's decision could not be regarded as having good and bad parts and would not be dismantled and reconstructed by the court.

RJT Consulting Engineers Ltd v *D M Engineering (NI) Ltd*, 9/5/2001 (Liverpool TCC)

An issue went to adjudication and the question later arose as to whether the oral agreement between the Parties was in 'writing' as far as the Construction Act was concerned. The court decided that it was, because of the 'comparatively great' relevant documentary evidence between the Parties which supported the existence of such an agreement. (See Court of Appeal decision, 8/3/2002.)

Re A Company (no. 1299 of 2001), 15/5/2001(E)

It was decided that, in the absence of a valid Section 111 Notice and without the need to go to adjudication, a Party had an undisputed debt for the amount asked for in an application for payment and could make a statutory demand which could lead to a petition to wind up a company. The company concerned alleged that it had matters to set off or abate, but had not taken any steps to pursue these. It therefore failed in its application for an injunction to restrain the presentation of a winding-up petition.

Faithful & Gould Ltd v *Arcal Ltd*, 25/5/2001 (E)

An Adjudicator was appointed by a company which later went into receivership. The court stated that the receiver had to pay the Adjudicator's fees. Further, it was confirmed that an Adjudicator was not obliged to sue for his fees as a natural person acting in his personal capacity.

Fence Gate Ltd v *James R Knowles Ltd*, 31/5/2001 (E)

Knowles submitted invoices for providing assistance in an arbitration and then invoices for litigation assistance. It was held that disputes in relation to fees for these services were not 'construction operations', even if the dispute to which they related was a 'construction operation'.

Barr Ltd v *Law Mining Ltd*, 15/6/2001 (S)

This Scottish case covers several important issues. It concerned a trial on preliminary issues arising from two separate adjudication decisions. It was decided that an Adjudicator has the jurisdiction to consider applications for payment that have not been certified and hence are arguably not 'due'.

It was not necessarily the case that the courts would always support an Adjudicator's decision where there had been a mistaken factual legal conclusion or a procedural error (*Sherwood & Casson* v *McKenzie*). The decision could be unsound but valid, or unsound but not valid because an Adjudicator had not the authority to make it. Also, a procedural error might take an Adjudicator outside its jurisdiction.

The issues were quite complex, there being two separate disputes and each contained several issues. The court said that it was for Adjudicators to decide at first instance whether what was at issue was one dispute or several and it was necessary to be realistic about this.

An Adjudicator cannot continue with issues arising from matters that occurred after contracts have been rescinded (unless the Adjudicator has been asked to decide whether the rescissions were lawful and decided that they were not) and it was not possible always to determine which sums were due from before and after the rescissions.

The court said that it was possible to separate good parts of the Adjudicator's decision from bad parts and granted a summary decree on a part on one adjudication decision and a full summary decree on the other. [*It will be noted that this is contrary to previous cases such as Farebrother.*]

Sindall Ltd v *Solland and Ors*, 15/6/2001 (E)

The court decided that an Adjudicator could decide a dispute about the underlying basis for a purported determination of a contract.

C & B Scene Concept Design Ltd v *Isobars Ltd (1)*, 20/6/2001 (E)

The Parties did not state how interim payments were to be made (ie at predetermined stages or at predetermined periods of time) in Clause 30 of a JCT Form with Contract's Design contract. A dispute arose and the Adjudicator applied some parts of Clause 30 to decide that, as written withholding notices were not provided as required, the Employer had to pay the full amount applied for by the Contractor in an interim application.

The Employer took the matter to court as it considered that:

(a) As the Parties did not state in Clause 30 how interim payments were to be made, then the whole of Clause 30 "fell away" and the payment provisions of the Scheme applied

(b) Under the Scheme (and the Construction Act) failure to give proper notices did not preclude making appropriate abatements to the amount applied for in determining an amount due

(c) Because of its failure to appreciate (a), the Adjudicator had addressed the wrong question and thus exceeded its jurisdiction

The Recorder agreed with (a) but considered (b) and (c) were arguable and so would not give summary judgement. (See Court of Appeal decision, 14/12/2001).

Ballast plc v *The Burrell Company (Construction Management) Ltd*, 21/6/2001 (S)

An Adjudicator concluded that in the circumstances of a case no adjudication was possible and issued a decision with the words 'not valid'. The court said this was a failure on the part of the Adjudicator to exercise his jurisdiction.

The court in this case also embarked upon a wide-ranging examination of the purpose of adjudication in Scotland and the circumstances in which a jurisdictional challenge could be made in Scotland.

William Naylor (t/a Powerfloated Concrete Floors) v *Greenacres Curling Ltd*, 26/6/2001 (S)

It was alleged that the Adjudicator had considered in a Scheme adjudication a dispute that had been decided in an earlier adjudication, which is not permitted by the Scheme. One of the Parties sought interdict and suspension (the equivalent of an injunction) against the Adjudicator. For reasons to do with the technicalities of Scottish practice, the Party chose the wrong court procedure and the interdict was recalled.

British Waterways Board, Petitioner (Judicial Review), 5/7/2001 (S)

In relation to a threatened adjudication under an NEC contract where Option Y(UK) 2 had been incorporated, the British Waterways Board raised a Petition for Judicial Review and sought interim orders on the basis that there was no 'dispute' in existence that carried an entitlement to go to adjudication. The other Party essentially considered that a provision in the contract that purported to redefine what a dispute was could be disregarded, as it was not compliant with the Construction Act.

The court decided that both the Act and a substituted Clause 90 envisaged 'the desirability of proceeding to a speedy resolution in matters in issue'. To create a dispute under the contract would mean a four-week delay. Under the circumstances the court considered that there was little additional preparation cost whether an adjudication proceeded immediately or four weeks later. On the balance of convenience, the interim orders were refused and the adjudication allowed to proceed.

City Inn Ltd v *Shepherd Construction Ltd*, 17/7/2001 (S)

The Parties entered into a contract that provided that any disputes would be resolved by adjudication and litigation in Scotland. An adjudication took place and a decision providing an extension of time to Shepherd was issued.

Notwithstanding the Adjudicator's award, an action relating to the payment of liquidated damages over the period and the repayment of loss and expenses was commenced by City Inn in the Court of Session in Scotland.

One of the issues was the burden of proof in such a case. Shepherd argued that since the contract provided 'the decision of the Adjudicator shall be binding on the Parties until the dispute or difference is finally determined by [litigation]' that meant that the Adjudicator's award stood until overturned and the onus or burden of proof in overturning it rested with City Inn. City Inn argued that such court proceedings would not be an appeal against the decision of the Adjudicator but were a reconsideration of the dispute as if no decision had been made by the Adjudicator. The court agreed with the latter view.

Gibson Lea Retail Interiors Ltd v *Macro Self-Service Wholesalers Ltd*, 24/7/2001 (E)

On a technical interpretation of the words 'forming part of the land' (which is a requirement for statutory adjudication), the court held that shop fitting was not a 'construction operation' and did not, therefore, carry an entitlement to adjudication under the Construction Act.

David McLean Housing Contractors Ltd v *Swansea Housing Association Ltd*, 27/7/2001, (TCC)(E)

This involved a JCT 81 contract. Practical completion had been certified and the Contractor submitted an application for payment which included the separate issues of loss and expense, valuation of variations. valuation of measured works, adjustment of provisional sums, and extensions of time.

Payment was not forthcoming, the Contractor commenced adjudication and an adjudication decision was reached on all issues, including the extension of time due. The Contractor and Employer made separate application to the courts, the former for the amount stated in the Adjudicator's decision, the latter for liquidated damages based on the end date which could be fixed from the extension of time included in the decision.

The courts agreed to summary judgement on both issues and so the Contractor, in fact, only received the balance due to it.

Another point decided by the courts was that the issues of extension of time, loss and expense and various valuation issues were all part of one dispute, ie all related to this one issue of the amount of money due to the Contractor. The courts said that if there had been two disputes these should have been the subject of two separate adjudications and the Adjudicator's appointment in this case would have been invalid.

SL Timber Systems Ltd v *Carillion Construction Ltd*, 27/7/2001 (S)

In this case the Adjudicator had decided that, as Carillion had not served Section 110 and 111 Notices, the full amount applied for in SLT's application for payment by SLT was due. Carillion refused to pay.

The court decided that firstly, in Scotland [*like England – see LPL*] the failure to issue a Section 110 Notice attracted no sanction, i.e. does not stop a Party disputing an amount claimed.

The court secondly confirmed that where a claim was made for the 'sums due under the Contract' this was not [*presumably necessarily*] the same as a sum claimed in an application and a dispute about whether the work in respect of which the claim was made had actually been done or had been properly measured or valued could still be considered by an Adjudicator in the absence of a withholding notice.

The court therefore found that the Adjudicator had erred but nevertheless acted within jurisdiction. It therefore made an order enforcing the award.

In what is perhaps a departure of Scottish procedure from English procedure, it was held that even in the absence of a valid Section 111 Notice, the claiming Party still had to prove any sum claimed in an adjudication.

Yarm Road Ltd v *Costain Ltd*, 30/7/2001 (E)

An original contract dated before the Construction Act came into force was made the subject of a novation agreement after the Act had come into force. The court held that the novation agreement fell within Section 104 of the Construction Act and that an adjudication was valid.

Millers Specialist Joinery Company Ltd v *Nobles Construction Ltd*, 3/8/2001 (E)

A Party went straight for court summary judgment on several outstanding invoices, without any intervening adjudication, on the grounds that in the absence of a valid Section 111 Notice there was no real prospect of a successful defence. The court awarded summary judgment on that basis, although reservations were expressed as to whether the claim itself would, in fact, have succeeded on its own merits.

Maxi Construction Management Ltd v Morton Rolls Ltd, 7/8/2001 (S)

A Party purported to make an application for payment under what they called 'Application Number 10'. However, the court decided that that it was not an application for payment at all but was simply an application for agreement of the Contractor's valuation. In any event, as it did not specify the basis on which it had been calculated, the purported notice did not comply with Paragraph 12 of the Scheme in Scotland.

Stubbs Rich Architect v *W H Tolley & Son Ltd*, 8/8/2001, (E)

An Adjudicator charged a fee of £1,561.50 which the respondent paid but considered it excessive. A claim to that effect was lodged in the Small Claims Court where the fee was agreed to be excessive. On appeal, a Recorder held that the fee was an integral part of the adjudication decision and in terms of Section 108 (4) of the Construction Act it enjoyed a similar immunity (except in bad faith).

Paul Jenson Ltd v *Stavely Industries plc*, 27/9/2001, (E)

The Parties asked an Adjudicator to decide on its own jurisdiction. The Adjudicator did so, deciding it did not have that jurisdiction.

On appeal, the Wigan County Court District Judge decided that the Parties had asked the Adjudicator to decide a preliminary issue on jurisdiction, the Adjudicator had done so, and that it was irrelevant whether the Adjudicator was correct or incorrect. It was also held that even if the Adjudicator was wrong, that did not amount to misconduct by the Adjudicator, and it was still entitled to such reasonable amounts as it might determine by way of fees and expenses.

William & Davis Oakley v *Airclear Environmental Ltd & TS Ltd*, 4/10/2001, (E).

The Parties sought to appoint an Adjudicator under the NAM/T and NAM/SC contracts. There was disagreement between the Parties as to the terms of the contracts, but an Adjudicator was appointed and an award issued which the Contractor sought to enforce by statutory demand.

An application to have the statutory demand set aside was initially dismissed by the courts. However, on appeal it was decided that the Adjudicator had not been properly appointed so that its decision was a nullity and the statutory demand fell away.

Jerome Engineering Ltd v *Lloyds Morris Electrical Ltd*, 23/11/2001, (E).

An Adjudicator was appointed under DOM2. The Adjudicator decided it could not reach a final value for variations but awarded an interim payment of £70,000, as well as the Adjudicator's fees. Payment not being forthcoming, proceedings were commenced in which the respondents argued that the relief sought was not specified in the Notice of Adjudication and therefore could not be made good by the Referral Notice. The court decided that Clause 38A.4.1 of DOM2 specifically anticipated the relief sought being specified in the Referral Notice and that in any event the Adjudicator had asked itself the correct question. The court upheld the decision, allowing interest on the Adjudicator's fee, but carrying forward interest on the interim award to any final account or arbitration.

Schimizu Europe Ltd v *Automajor Ltd*, 17/1/2002, (TCC) (E)

An Adjudicator was appointed under a design and build contract. One of the matters under dispute involved Schimizu alterations to smoke ventilation works. Automajor paid, without prejudice, £50,000 towards those costs. The Adjudicator decided that the alterations to the smoke ventilation works were not variations, and further, that it had no jurisdiction to factor into its calculations the £50,000. Automajor made payment, with the exception of that part of the award relating to the smoke ventilation work.

Schimizu sought enforcement of the decision on the latter. The court decided that by making part-payment on the decision, Automajor had waived any objections it might have and had elected to treat the decision as valid, that this was not the sort of adjudication decision where Parties could pick and choose which parts should be enforceable or unenforceable, and finally, since the Adjudicator had been asked to determine the sums due, any mistakes in that respect did not go to its jurisdiction.

C&B Scene Concepts Design Ltd v *Isobars Ltd*, 14/12/2001. (Court of Appeal) (E)

The situation is as stated in the summary given against this case when it first arose on 20/6/2001 and will not be repeated here.

The Appeal Court would only decide on point (c). It started from the assumption that the Employer's contentions for (a) and (b) were correct and decided that the Adjudicator had not gone beyond the actual issues it had been asked to consider and that even if it had made errors in law relating to the relevant contractual provisions, its decision was still binding.

Karl Construction (Scotland) Ltd v *Sweeney Civil Engineering (Scotland) Ltd*, 17/1/2002, (Court of Session) (S)

An Extra Decision of the Inner House of the Court of Session supported the original decision of 21/12/2000.

Solland International v *Daraydon Holdings*, 15/2/2002, (S)

There was no dispute over an Adjudicator's decision on the valuation of an interim application but Daraydon sought to deduct substantial liquidated damages from the amount due. The courts did not agree to this, stating that the amount decided by the Adjudicator should be paid.

[This issue relates to supporting an adjudication decision, there being no separate application for summary judgement on the liquidated damages as was the case in McLean *v* Swansea Housing. *Contractually the Employer was still entitled to deduct liquidated damages from subsequent interim valuations, or simply to ask for that money if there were no further valuations and then to go to adjudication if payment was not forthcoming.]*

Total M & E Services Ltd v *ABB Building Technologies Ltd (formerly ABB Stewards Ltd)*, 26/2/2002, (TCC) (E)

It was confirmed that as there was no provision for the recovery of adjudication costs in the Construction Act or elsewhere in the pertaining documents, costs could not be recovered as damages.

RJT Consulting Engineers Ltd v *D M Engineering (NI) Ltd*, 8/3/2002, (Court of Appeal) (E)

This was an appeal against an earlier decision given at the Liverpool TCC on 9/5/2001. It was a unanimous decision by the three judges concerned that the original court decision was incorrect, ie that the large amount of available documentary evidence did not amount to adequate evidence of an agreement in writing for the purposes of the Construction Act.

The Court agreed an earlier judgement which states:

> *Disputes as to the terms, express and implied, of oral construction agreements are not readily susceptible of resolution by a summary procedure such as adjudication.*

Other statements included:

1. *The written record of the agreement is the foundation from which a dispute may spring … the least the Adjudicator has to be certain about is the terms of the agreement which is giving rise to this dispute.*
2. *The Adjudicator has to start with some certainty as to what the terms of the contract are.*
3. *Parliament evidently decided that it was inappropriate for an Adjudicator to have to deal with disputes which often arise as to the terms of an oral agreement.*

Two appeal judges agreed that '*what has to be evidenced in writing is, literally, the agreement, which means all of it, not part of it …. The only exception …. is …. where the material or relevant parts alleged and not denied in the written submissions in the adjudication proceedings are sufficient.*'

The third appeal judge, although agreeing in this case that the appeal should be allowed, said that *'What is important is that the terms of the agreement material to the issue giving rise to the reference should be clearly recorded in writing, not ... every term, however trivial and unrelated to those issued ....'*

Edmund Nuttall Limited v *R G Carter Limited*, 21/3/2002, EWHC 400, (TCC) (E)

This case involved a DOM/1 standard form of sub-contract. There was an initial claim for May 2001 for amounts under 9 headings, supported by some evidence. It was rejected as not being adequately substantiated. After some further discussion it was submitted to adjudication, with some items being dropped and the others substantiated by significant additional documentation, much of which had not been seen until then by the Responding Party. They protested, saying that in consequence there was not yet a dispute and hence the Adjudicator had no jurisdiction. The Adjudicator nevertheless proceeded to a decision, which was appealed on the basis of lack of jurisdiction.

Various case decisions were quoted by the Parties and by the judge in its decision, including *Fastrack* v *Morrison*, *Monmouthshire* v *Costelloe*, *Halki Shipping* v *Sopex*, *Sindall* v *Solland*, *K & D Construction* v *Midas Homes*. All in all, it was a very detailed analysis of what constitutes a claim and what constitutes a dispute.

The court stated that the key question was whether the Adjudicator had adjudicated on the dispute described in the Notice of Referral dated 14 December 2001 or on another dispute.

The courts decided that what was referred to adjudication on 14 December 2001 was the May claim 'comprising the package of facts relied upon [by] each side in support of the[ir] respective positions ... and the arguments which had been rehearsed.'

The judge considered that the new information provided by the Referring Party was not part of the May claim and hence the Adjudicator had decided on the basis of things that were not referred to it. There was therefore no jurisdiction for the decision and it was unenforceable.

[This is a clear example of where adjudication is not the same as arbitration or litigation (where the Parties can more freely submit new information and Particulars to the tribunal) and Adjudicators (or legal representatives) should not attempt to try to make it so. It should be noted that a significant extract from this key judgement is included verbatim for ease of reference in the main text under the heading 'What is a dispute' in Chapter 14.]

Balfour Beatty Construction Ltd v *The Mayor and Burgesses of the London Borough of Lambeth*, 12/4/2002, Bliss EWHC 597 (TCC) (E)

This case involved a JCT 1998 Standard Form of Building Contract, Local Authorities without Quantities, incorporating TC/94 and Contractor's Designed Portion Supplement 1998. The dispute related to disagreement on an extension of time and associated damages that were deducted because of delay. The contractor's delay claim was not properly substantiated and was therefore rejected by the Employer. The Adjudicator asked for more information and undertook its own delay analysis, leading to an award of an extension of time.

The decision was appealed under several headings, including with the rules of natural justice, bias and non-compliance of the Adjudicator with the rules of natural justice as no opportunity had been given to the Parties to study and respond to its analysis.

Various case decisions relating to fairness and natural justice were quoted by the Parties and the judge in its decision, including some related to arbitration, but also including *Discain* v *Opecprime* 2000 and 2001.

The judge quoted from *Discain* v *Opecprime* 2000, which stated:

> *... the adjudicator is working under pressure of time and circumstance which makes it extremely difficult to comply with the rules of natural justice in the manner of a court or an arbitrator. ... the system created by the [HGCRA] can only be made to work in practice if some breaches of the rules of natural justice which have no demonstrable consequence are disregarded.*

In general, the judge in this case agreed that peripheral or irrelevant non-material issues do not need to be brought to the attention of both Parties, particularly because of the provisional nature of adjudication decisions, (although the judge did not comment on those adjudication decisions that might be agreed to be final and conclusive).

With regard to this case in particular, the judge noted that Balfour Beatty did little or nothing to present its case 'in a logical or methodical way' as they [apparently] decided that it was not practicable and, indeed, it was unnecessary to do so. Balfour Beatty could not even produce basic information in the time required when requested by the Adjudicator.

The judge continued

> *... the adjudicator not only took the initiative in ascertaining the facts but also applied his own knowledge and experience of them and thus, in effect, did BB's work for it ....* [He did not] *however inform Lambeth of what he then intended to do with the facts. He did not invite their comments ...* nor *inform either party of the methodology that he intended to adopt, or to seek observations from them as to the manner in which it or any other methodology might reasonably and properly be used in the circumstances to establish or to test BB's case. ... he ought to have done so. ... One would ordinarily expect the appropriate method of analysis to be agreed before it was used by an architect or other contract administrator. The adjudicator steps into the shoes of such a person. If an adjudicator intends to use a method which was not agreed and has not been put forward as appropriate by either party he ought to inform the parties and to obtain their views as it is his choice of how the dispute might be decided. An adjudicator is of course entitled to use the powers available to him but he may not of his own volition use them to make good fundamental deficiencies in the material presented by one party without first giving the other party an proper opportunity of dealing both with the intention and with the results.*

> *... Lambeth was entitled to criticise BB's case without putting forward an alternative. Since BB had not justified its case Lambeth was not obliged to justify the architect's extensions of time or certificates of non-completion. ... Lambeth was not obliged to prove that they were right (although it is usually prudent to do so). It was entitled to have the dispute decided on BB's own terms, i.e. on the material provided by it either originally or in answer to the adjudicator's requests and not on a basis devised by the adjudicator and which had not bee made known to it. For example, if a dispute about the ascertainment of loss or expense arose because the quantity surveyor considered that the contractor had not provided the necessary material, the adjudicator would have to decide that dispute on that basis, i.e. whether the loss and expense was capable of being ascertained on what was known to the quantity surveyor. ...* The adjudicator *exceeded his jurisdiction by himself making good fundamental deficiencies in BB's material, namely the lack of a critical path and the method of analysis adopted for demonstrating the criticality or otherwise of the Relevant Events.*

The judge then stated

> *constructing (or reconstructing) a party's case for it without confronting the other party with it is such a potentially serious breach of the requirement of either impartiality or fairness that the decision is invalid for it is not a decision which the adjudicator was authorised to make.*

And, elsewhere in the judgement, that

> *an observer would conclude that by making good the deficiencies in BB's case and by overcoming the absence of a sustainable as-built programme with a critical path and the complete lack of any analysis as to which of the relevant events were critical and non-critical, the adjudicator moved into the danger zone of being impartial or liable to 'apparent bias', as it is now recognised. That lack of impartially or supposed bias can easily be cured by disclosure to the other party of what is being done or thought about.*

Whilst the judge considered that, in this case, the Adjudicator had time to inform the Parties and invite comment on the 'as built' programme and critical path he had devised and on his intended method of analysis, the judge also gave some advice of general application. He stated that where the nature of a case 'was likely to make it extremely difficult' for an Adjudicator

> *to complete a reasonably fair investigation of it within the original 28 days, or even the further period agreed by the parties, then an adjudicator would have to say that it would not be possible to carry out such an investigation and to arrive at a decision, even if it was 'coarse' within the time available. An adjudicator does not act impartially or fairly if he arrives at a decision without having given a party a reasonable opportunity of commenting upon the case that it has to meet (whether presented by the other party or thought to be important by the adjudicator) simply because*

> *there is not enough time available. An adjudicator, acting impartially and in accordance with the principles of natural justice, ought in such circumstances to inform the parties that a decision could not properly reasonably and fairly be arrived at within the time and invite the parties to agree further time. If the parties were not able to agree more time then an adjudicator ought not to make a decision at all and should resign.*

[This case is fundamental to clarifying the court's position in several areas, particularly how an Adjudicator should tackle inadequately presented evidence, how delay issues should be tackled, the extent to which an Adjudicator can undertake its own analysis, natural justice requirements notwithstanding the limited timescale for a decision. For this reason the summary is more extensive than for previous ones. It will be noted that on the natural justice issue the criticism was not about gaining approval from the Parties – more a case of informing them about certain aspects and inviting comments which the Adjudicator should assess when deciding a course of action or making its decision.]

Bibliography

ANDERSON, R. N. M. *Adjudication under the NEC.* Thomas Telford, 2001.
— *A Practical Guide to Adjudication in Construction Matters.* W. Green/ Sweet & Maxwell, 2000.
ARBITRATION ACT, 1996
BEATSON, J. AND FRIEDMAN, D. (eds), *Good Faith and Fault in Contract Law.* Clarendon Press, 1995, p. 153.
BURN, E. H. *Cheshire and Burn's Modern Law of Real Property.* 15th edn, Butterworths, 1994.
CIC *The CIC Model Adjudication Procedure.* 2nd edn. Construction Industry Council, 1998.
CIC *The Guide To Construction Adjudication.* Construction Industry Council, 1998.
CONSTRUCTION (RIGHTS OF THIRD PARTIES) ACT, 1999.
CONSTRUCTION UMBRELLA BODIES' ADJUDICATION TASK GROUP DRAFT, *Adjudication under the Scheme – Guidance to Adjudicators.* 2 May 2002.
COTTAM, G. *Adjudication under the Scheme for Construction Contracts.* Thomas Telford, 1998.
FRANCIS, T. AND NIGHTINGALE, S. *Adjudication under the Construction Act.* Monitor Press, 1998.
GC/WORKS/1, Without Quantities. (1998)
GC/WORKS/1, Model Forms and Commentary. (1998)
GOFF, R. AND JONES, G. *The Law of Restitution*, 5th edn, Sweet and Maxwell, 1998.
HIBBERD, P. AND NEWMAN, P. *ADR and Adjudication in Construction Contracts.* Blackwell Science, 1999.
THE HOUSING GRANTS, CONSTRUCTION AND REGENERATION ACT. 1996.
ICE *Conditions of Contract.* 7th edn, Measurement Version, 1999.
ICE *Conditions of Contract.* 7th edn, Measurement Version, Guidance Notes, 1999.
ICE *Adjudication Procedure.* (1997), Thomas Telford, 1998.
JCT 80, *Amendment 18.* 1998.
JONES, B. L. *Garner's Administrative Law.* 7th edn, Butterworths, 1989. 'Grounds for Challenge' pp. 157–62.
KENDALL, J. *Expert Determination.* 3rd edn, Sweet and Maxwell, 2001.
MUSTILL, M. J. AND BOYD, S. C., *Commercial Arbitration.* 2nd edn, Butterworths, 1989.

REDMOND, J. *Adjudication in Construction Contracts.* Blackwell Science, 2001.

RICHES, J. AND DANCASTER, C. *Construction Adjudication – A Practical Guide.* LLP Limited, 1999.

THE SCHEME FOR CONSTRUCTION CONTRACTS (ENGLAND AND WALES) REGULATIONS. 1998.

THE SCHEME FOR CONSTRUCTION CONTRACTS (AMENDMENT) (ENGLAND) REGULATIONS. 2001.

STEVENSON, R. AND CHAPMAN, P. *Construction Adjudication.* Jordan Publishing Limited, 1999.

TIMPSON, J. AND TOTTERDILL, B. *Adjudication for Architects and Engineers.* Thomas Telford, 1999.

TOTTERDILL, B., *ICE Adjudication Procedure (1997) – A User's Guide and Commentary to the ICE Adjudication Procedure.* Thomas Telford, 1998.

WOOD, M. *The Construction Act – A Practical Guide.* Chandos Publishing, 1998.